ACKNOWLEDGEMENT

First and foremost, praises and thanks to Our Parents and Our Family Members, the God, the Almighty, and well-wishers for their showers of blessings throughout this book work and to complete the publication successfully.

The completion of this undertaking would not have been possible without the participation and assistance of many people whose names may not be enumerated here. Their contributions are sincerely appreciated and gratefully acknowledged. However, the group likes to express their deep appreciation and indebtedness particularly to the following:

We are pleased to express Our deepest sense of gratitude to Dr. Madan A Sendhil Chairman, Rathinam Group, Coimbatore, for providing a chance to bring out this book and good wishes to him.

We render Our sincere thanks to family members, relatives, friends and others who were very much supportive both morally and physically during my publication.

Above all, we render Our heartfelt gratitude and thanks to the Great Almighty, the author of knowledge and wisdom, for his abundant love.

PROGRAMMING IN C

(For Common to B.E CSE & IT Branches)

AS PER THE LATEST SYLLABUS OF ANNA UNIVERSITY, CHENNAI

(Regulation 2017)

Dr. K.SIVAKUMAR
Principal
Professor & Dean
Dept. of Computer Science and Engineering
Rathinam Technical Campus Coimbatore,
Tamil Nadu.

Mr. D.KALEESWARAN
Assistant Professor
Dept. of Computer Science and Engineering
Rathinam Technical Campus Coimbatore,
Tamil Nadu.

SHANLAX PUBLICATIONS

PROGRAMMING IN C

First Edition: September, 2018

ISBN: 978-81-922811-6-2

Second Edition: July, 2020

ISBN: 978-93-90082-28-5

Price: Rs.330/-

Publisher
SHANLAX PUBLICATIONS
61, 66 T.P.K. Main Road
Vasantha Nagar
Madurai – 625003
Tamil Nadu, India

Ph: 0452-4208765,
Mobile: 7639303383
email:publisher@shanlaxpublications.com
web: www.shanlaxpublications.com

ANNA UNIVERSITY CHENNAI – REGULATION 2017
PROGRAMMING IN C

UNIT I BASICS OF C PROGRAMMING

Introduction to programming paradigms - Structure of C program – C programming: Data Types – Storage classes - Constants – Enumeration Constants - Keywords – Operators: Precedence and Associativity - Expressions - Input/Output statements, Assignment statements – Decision making statements – Switch statement - Looping statements – Pre-processor directives - Compilation process.

UNIT II ARRAYS AND STRINGS

Introduction to Arrays: Declaration, Initialization – One dimensional array – Example Program: Computing Mean, Median and Mode - Two dimensional arrays – Example Program: Matrix Operations (Addition, Scaling, Determinant and Transpose) - String operations: length, compare, concatenate, copy – Selection sort, linear and binary search.

UNIT III FUNCTIONS AND POINTERS

Introduction to functions: Function prototype, function definition, function call, Built-in functions (string functions, math functions) – Recursion – Example Program: Computation of Sine series, Scientific calculator using built-in functions, Binary Search using recursive functions – Pointers – Pointer operators – Pointer arithmetic – Arrays and pointers – Array of pointers – Example Program: Sorting of names – Parameter passing: Pass by value, Pass by reference – Example Program: Swapping of two numbers and changing the value of a variable using pass by reference.

UNIT IV STRUCTURES

Structure - Nested structures - Pointer and Structures - Array of structures - Example Program using structures and pointers - Self referential structures - Dynamic memory allocation - Singly linked list - typedef.

UNIT V FILE PROCESSING

Files - Types of file processing: Sequential access, Random access - Sequential access file - Example Program: Finding average of numbers stored in sequential access file - Random access file - Example Program: Transaction processing using random access files - Command line arguments.

CONTENTS

UNIT I

BASICS OF C PROGRAMMING

Introduction to programming paradigms - Structure of C program - C programming: Data Types - Storage classes - Constants - Enumeration Constants - Keywords - Operators: Precedence and Associativity - Expressions - Input/Output statements, Assignment statements - Decision making statements - Switch statement - Looping statements - Pre-processor directives - Compilation process

1.1 Introduction to Programming Paradigms

- A programming paradigm is a way of programming, a style of solving problems and thinking about the domain. Languages directly or indirectly influence programming style; for example, Haskell is purely functional while Python allows a blend of OOP (Object Oriented Programming), functional and imperative programming.
- Even though most new languages are multi-paradigm, they usually make it easiest to program in a single paradigm; C programs can be written in a functional programming style (instead of the orthodox procedural fashion) – it is just extremely difficult. Ever wondered why classes are needed to add two numbers in OOP Java?
- Paradigms can be viewed as abstractions and metaphors for modeling problems – logic programs can be viewed as theorem solvers; functional programs rely on mathematical models while OOP models real-life objects and their interactions.
- Nearly every standard problem can be solved using any of the programming paradigms however some problems are best solved using some paradigms (e.g. recursion vs loops). Thus knowing many paradigms helps programmers in producing very neat and elegant solutions; since they know different ways of solving problems and modeling domains. Let‘s take a shallow dive into some of the popular paradigms.

1.1.1. Imperative programming

Imperate: *Done by express direction, not involuntary; commanded*

- In the imperative paradigm, programs are written as a series of instructions that affect global state. It is conceptually similar to cooking recipes which strictly list the steps in preparing a dish.
- The imperative paradigm was quite popular until Dijkstra published his famous _GOTO considered harmful_ letter. This fueled the adoption of the structured programming paradigm (albeit at the detriment of imperative programming).

In the itemized list below we describe the main properties of the imperative paradigm.

Characteristics

- Discipline and idea
  - Digital hardware technology and the ideas of Von Neumann
- Incremental change of the program state as a function of time.
- Execution of computational steps in an order governed by control structures
  - We call the steps for commands
- Straightforward abstractions of the way a traditional Von Neumann computer works
- Similar to descriptions of everyday routines, such as food recipes and car repair
- Typical commands offered by imperative languages
  - Assignment, IO, procedure calls
- Language representatives
  - Fortran, Algol, Pascal, Basic, C

1.1.1. Structured programming

- Structured programming is somewhat similar to imperative programming however control flow is defined in terms of loops, subroutines and conditionals; this distinguishes it from the imperative model of jumping around at the heck and bequest of every GOTO call.
- The introduction of lexical scoping also helps to prevent the inadvertent overwriting of variables in the global scope.

Characteristics

- Complexity can be reduced using the concepts of divide and conquer.
- Logical structures ensure clear flow of control.
- Increase in productivity by allowing multiple programmers to work on different parts of the project independently at the same time.
- Modules can be re-used many times, thus it saves time, reduces complexity and increase reliability.
- Easier to update/fix the program by replacing individual modules rather than larger amount of code.
- Ability to either eliminate or at least reduce the necessity of employing GOTO statement.

1.1.2. Declarative programming

- Declarative programs have implicit execution: programs say what they want but not how the results should be gotten. SQL is a popular example, have you ever wondered how the SQL engine is able to calculate sums, averages, max etc.
- A conceptual example is that of a manager declaring sales targets for his marketing team: the team is smart enough to figure out heuristics and also to go do it on their own (hopefully within ethical and legal limits).

Characteristics

- Model of computation based on a system where relationships are specified directly in terms of the constituents of the input data.
- Made up of sets of definitions or equations describing relations which specify what is to be computed, not how it is to be computed.
- Non-destructive assignment of variables.
- Explicit representations for data structures used.
- Order of execution does not matter (no side effects).
- Expressions/definitions can be used as values.
- Programmer no longer responsible for control.

1.1.3. Functional Programming (FP)

- The functional programming paradigm is a subset of the declarative paradigm. All computation is done using functions – it attempts to do away with the use of variables and value assignments.
- Pure functional programs are written as a combination of function calls which have no side effects and return the same result every time they are called with the same input no matter the environment. The elegance of FP comes from the use of higher-order functions (functions which can accept/ return other functions e.g. differentiation/ integration in the maths domain), referential transparency and high levels of abstraction.
- On the other hand, some concepts might be inefficient (e.g. Y combinator) or ugly.

Below we characterize the most important, overall properties of the functional programming paradigm.

Characteristics

- Discipline and idea
    - Mathematics and the theory of functions
- The values produced are non-mutable
    - Impossible to change any constituent of a composite value
    - As a remedy, it is possible to make a revised copy of composite value
- A temporal
    - Abstracts a single expression to a function which can be evaluated as an expression
- Functions are first class values
    - Functions are full-fledged data just like numbers, lists, ...
- Fits well with computations driven by needs
    - Opens a new world of possibilities

1.1.4. Object Oriented Programming (OOP)

- This is probably the most popular and probably most widely-used of all programming paradigms. Object oriented programming is built on four major principles – encapsulation, polymorphism, inheritance and data abstraction.

- OOP allows the creation of abstract models that mirror real-life objects' interactions and behaviour. This enables the creation of huge complex software systems having intricate interactions such as state preservation, communication exchanges and behaviour handling.
- There are two flavours: **classical** (where objects are defined and share their characteristics and behaviour based on a class' blueprint) and **prototypical** (objects get their behaviour and characteristics from an original prototype object).

As for the other main programming paradigms, we will now describe the most important properties of object-oriented programming.

Characteristics

- Discipline and idea
  - The theory of concepts, and models of human interaction with real world
- phenomena
- Data as well as operations are encapsulated in objects
- Information hiding is used to protect internal properties of an object
- Objects interact by means of message passing
  - A metaphor for applying an operation on an object
- In most object-oriented languages objects are grouped in classes
  - Objects in classes are similar enough to allow programming of the classes, as opposed to programming of the individual objects
  - Classes represent concepts whereas objects represent phenomena
- Classes are organized in inheritance hierarchies
  - Provides for class extension or specialization

1.1.5. Conclusion

- Some other paradigms include the event-driven where program respond to events (think web apps) while the logical programming paradigm tries to infer the solution when it is given a set of constraints and or conditions.

- Upcoming posts would insha Allaah take a more thorough look at some of the popular paradigms.

1.2. Structure of C Program

A C program basically consists of the following parts:

- Preprocessor Commands
- Functions
- Variables
- Statements & Expressions
- Comments

Let us look at a simple code that would print the words "Hello World".

```
#include <stdio.h>
int main()
{
/* my first program in C */
printf("Hello, World! \n");
return 0;
}
```

A C program source code can be written in any text editor; however the file should be saved with .c extension.

Let us take a look at the various parts of the above program –

- The first line of the program #include <stdio.h> is a preprocessor command, which tells a C compiler to include stdio.h file before going to actual compilation.
- The next line int main() is the main function where the program execution begins.
- The next line /*...*/ will be ignored by the compiler and it has been put to add additional comments in the program. So such lines are called comments in the program.
- The next line printf(...) is another function available in C which causes the message "Hello, World!" to be displayed on the screen.

- The next line return 0; terminates the main() function and returns the value 0.

1.2.1.1. Executing a C program in command prompt using Turbo C compiler

The Turbo C compiler *TCC* is existed within the *bin* directory of Turbo C default folder *TC* (c:\tc\bin). It can be accesses only within the *bin* directory so that, C programs in other locations can't be compiled. To access The Turbo C compiler *TCC* from anywhere, we set the path to its parent folder c:\tc*bin*.

By setting path we inform the operating system to show files of a particular folder to anywhere.

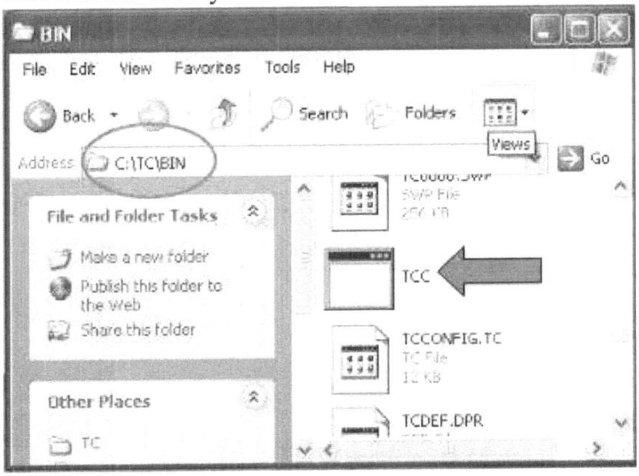

Setting path to the parent folder of TCC (c:\tc\bin)

Step 1: Right click on –My computer‖ – select –properties‖

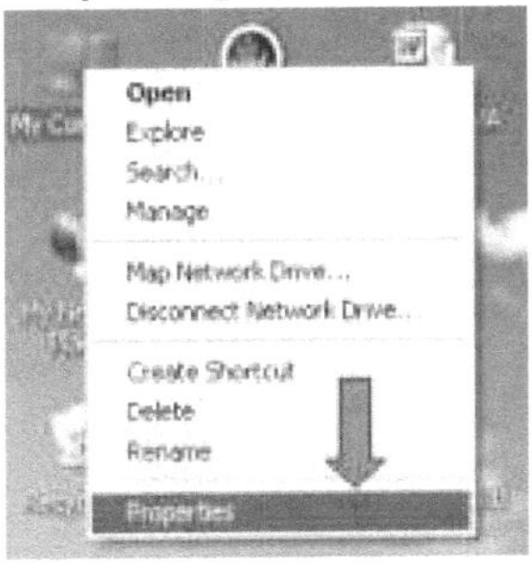

Step 2: select —advanced‖ tab – select —Environment Variables‖

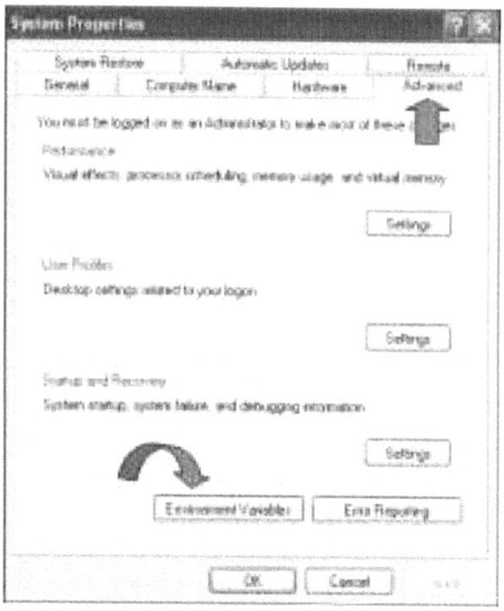

Step 3: Select —path‖ variable – select edit

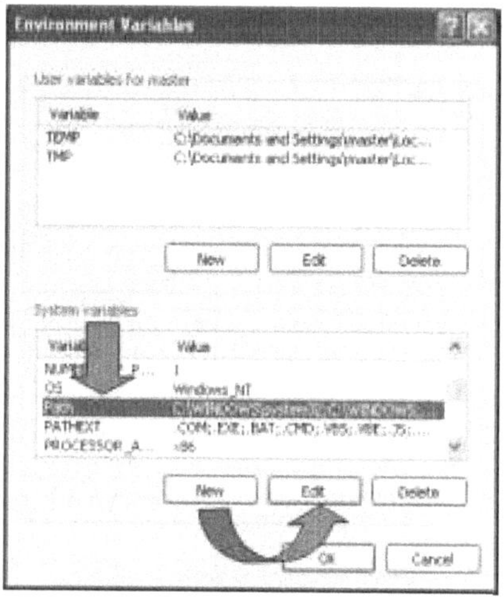

Step 4: Add *c:\tc\bin;.;* to the path value. Make sure that the existed path must not be erased, which result some of the applications may malfunction.

Executing the C program

Step1: Go to the command prompt by selecting start – run – type cmd – select open

Step2: create your personal folder (demo) and change to it.

Step3: open the text editor with the program name (first.c)

```
c:\....>d: d:\>md
demo d:\>cd demo
d:\demo>edit first.c
```

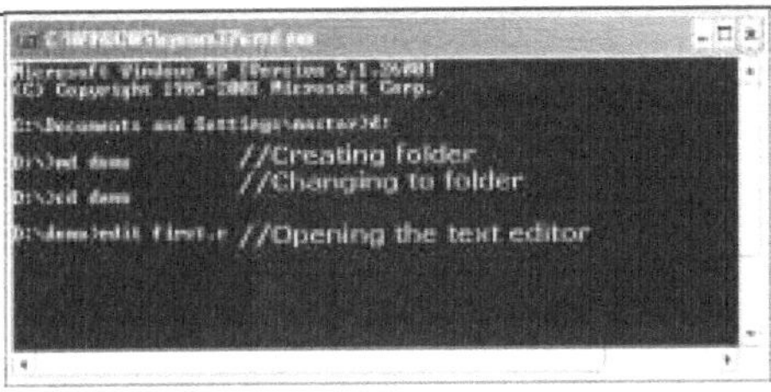

Step4: Type the program

```
#include<stdio.h>
int main()
{
printf("Hello World");
return 0;
}
```

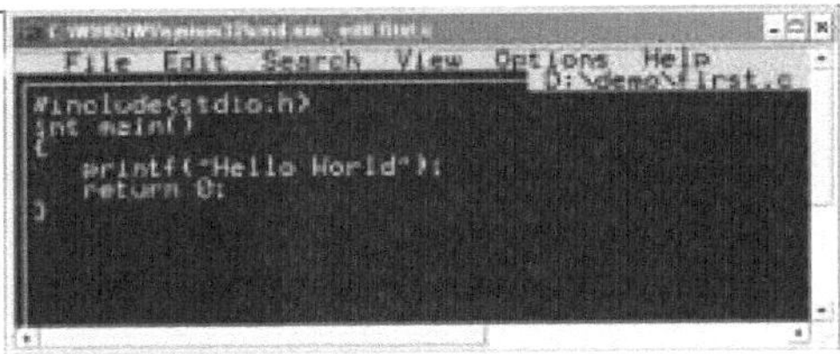

Step5: Save and exit from the text editor by selecting file – exit

Step6: compile the program (first.c) using the turbo C compiler TCC, It produces the executable file with the program name (first.exe)

Step7: execute the executable file with the file name.

```
d:\demo>tcc first.c
d:\demo>first
```

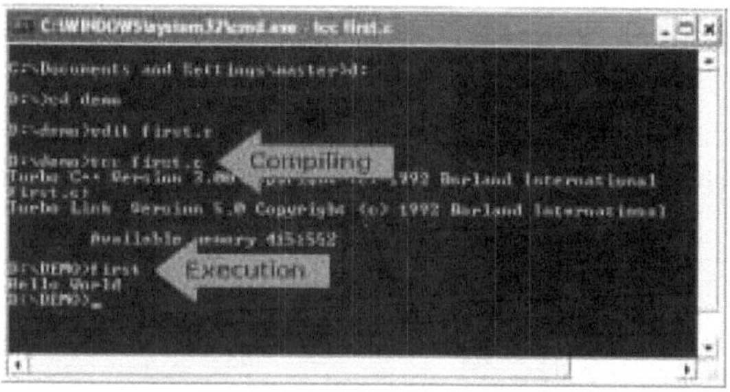

1.3. C Programming: Data Types

You will learn about data types and how to declare a variable in C programming.

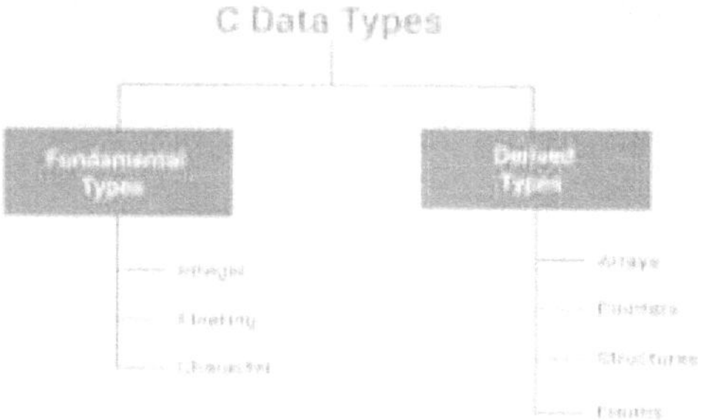

Figure1: C Programming: Data Types

Each variable in C has an associated data type. Each data type requires different amounts of memory and has some specific operations which can be performed over it. Let us briefly describe them one by one:

Following are the examples of some very common data types used in C:

- **char:** The most basic data type in C. It stores a single character and requires a single byte of memory in almost all compilers.
- **int:** As the name suggests, an int variable is used to store an integer.
- **float:** It is used to store decimal numbers (numbers with floating point value) with single precision.
- **double:** It is used to store decimal numbers (numbers with floating point value) with double precision.

Different data types also have different ranges upto which they can store numbers. These ranges may vary from compiler to compiler. Below is list of ranges along with the memory requirement and format specifiers on 32 bit gcc compiler.

Data Type	**Memory (bytes)**	**Range**	**Format Specifier**
short int	2	-32,768 to 32,767	%hd
unsigned short int	2	0 to 65,535	%hu
unsigned int	4	0 to 4,294,967,295	%u
int	4	-2,147,483,648 to 2,147,483,647	%d
long int	4	-2,147,483,648 to 2,147,483,647	%ld
unsigned long int	4	0 to 4,294,967,295	%lu
long long int	8	-(2^63) to (2^63)-1	%lld
unsigned long long int	8	0 to 18,446,744,073,709,551,615	%llu
signed char	1	-128 to 127	%c
unsigned char	1	0 to 255	%c
float	4		%f
double	8		%lf
long double	12		%Lf

See the following C program for the usage of the various data types

```
#include <stdio.h>
int main()
{
int a = 1; char b
='G'; double c =
3.14;
printf("Hello World!\n");

//printing the variables defined above along with their sizes
printf("Hello! I am a character. My value is %c and "
```

```
"my size is %lu byte.\n", b,sizeof(char));
//can use sizeof(b) above as well

printf("Hello! I am an integer. My value is %d and "
 "my size is %lubytes.\n", a,sizeof(int));
//can use sizeof(a) above as well

printf("Hello! I am a double floating point variable."
 " My value is %lf and my size is %lu bytes.\n",c,sizeof(double));
//can use sizeof(c) above as well

printf("Bye! See you soon. :)\n");

return 0;
}
```

Output

```
Hello World!
Hello! I am a character. My value is G and my size is 1 byte.
Hello! I am an integer. My value is 1 and my size is 4bytes.
Hello! I am a double floating point variable. My value is 3.140000
and my size is 8 bytes.
Bye! See you soon. :)
```

1.3.1. sizeof() function in C language

The **sizeof()** function is used to find the memory space allocated for each C data types.

```
#include <stdio.h>
#include <limits.h>
int main()
{
 int a; char
 b; float c;
 double d;
 printf("Storage size for int data type:%d \n",sizeof(a));
 printf("Storage size for char data type:%d \n",sizeof(b));
```

```
 printf("Storage size for float data type:%d \n",sizeof(c));
 printf("Storage size for double data type:%d\n",sizeof(d));
 return 0;
}
```

Output:

```
Storage size for int data type:4
Storage size for char data type:1
Storage size for float data type:4
Storage size for double data type:8
```

1.3.2. Modifiers in C language

- The amount of memory space to be allocated for a variable is derived by modifiers.
- Modifiers are prefixed with basic data types to modify (either increase or decrease) the amount of storage space allocated to a variable.
- For example, storage space for int data type is 4 byte for 32 bit processor. We can increase the range by using long int which is 8 byte. We can decrease the range by using short int which is 2 byte.

There are 5 modifiers available in C language. They are:

- short
- long
- signed
- unsigned
- long long

1.3.3. Enumeration data type in C language

- Enumeration data type consists of named integer constants as a list.
- It start with 0 (zero) by default and value is incremented by 1 for the sequential identifiers in the list.

Enum syntax in C

```
enum identifier [optional{ enumerator-list }];
```

Enum example in C

- enum month { Jan, Feb, Mar }; or

/* Jan, Feb and Mar variables will be assigned to 0, 1 and 2 respectively by default */

- enum month { Jan = 1, Feb, Mar };

/* Feb and Mar variables will be assigned to 2 and 3 respectively by default */

- enum month { Jan = 20, Feb, Mar };

/* Jan is assigned to 20. Feb and Mar variables will be assigned to 21 and 22 respectively by default */

The above enum functionality can also be implemented by ―#define‖ preprocessor directive as given below. Above enum example is same as given below.

```
#define Jan 20;
#define Feb 21;
#define Mar 22;
```

Example

```
#include <stdio.h>
int main()
{
 enum MONTH { Jan = 0, Feb, Mar };
 enum MONTH month = Mar;
 if(month == 0)
 printf("Value of Jan");
 else if(month == 1)
 printf("Month is Feb");
 if(month == 2)
 printf("Month is Mar");
}
```

Output

```
Month is Mar
```

1.3.4. Derived data type in C language

- Array, pointer, structure and union are called derived data type in C language.

1.3.5. Void data type in C language

- Void is an empty data type that has no value.
- This can be used in functions and pointers

1.4. Storage Classes in C Programming

- A storage class defines the scope (visibility) and life-time of variables and/or functions within a C Program.
- They precede the type that they modify.

We have four different storage classes in a C program –

- auto
- register
- static
- extern

The auto Storage Class

The auto storage class is the default storage class for all local variables.

```
{
int mount;
auto int month;
}
```

The example above defines two variables with in the same storage class. 'auto' can only be used within functions, i.e., local variables.

The register Storage Class

The register storage class is used to define local variables that should be stored in a register instead of RAM. This means that the variable has a maximum size equal to the register size (usually one word) and can't have the unary '&' operator applied to it (as it does not have a memory location).

```
{
register intmiles;
}
```

The register should only be used for variables that require quick access such as counters. It should also be noted that defining 'register' does not mean that the variable will be stored in a register. It means that it MIGHT be stored in a register depending on hardware and implementation restrictions.

The static Storage Class

The static storage class instructs the compiler to keep a local variable in existence during the life-time of the program instead of creating and destroying it each time it comes into and goes out of scope. Therefore, making local variables static allows them to maintain their values between function calls.

The static modifier may also be applied to global variables. When this is done, it causes that variable's scope to be restricted to the file in which it is declared.

In C programming, when static is used on a global variable, it causes only one copy of that member to be shared by all the objects of its class.

```
#include <stdio.h>
/* function declaration */
void func(void);
static int count = 5; /* global variable */
 main()
{
 while(count--) {
func();
 }
 return 0;
}
/* function definition */
void func( void ) {
 static int i = 5; /* local static variable */
 i++;
 printf("i is %d and count is %d\n", i, count);
}
```

When the above code is compiled and executed, it produces the following result .

```
i is 6 and count is 4
i is 7 and count is 3
i is 8 and count is 2
i is 9 and count is 1
i is 10 and count is 0
```

The extern Storage Class

The extern storage class is used to give a reference of a global variable that is visible to ALL the program files. When you use 'extern', the variable cannot be initialized however, it points the variable name at a storage location that has been previously defined.

When you have multiple files and you define a global variable or function, which will also be used in other files, then extern will be used in another file to provide the reference of defined variable or function. Just for understanding, extern is used to declare a global variable or function in another file.

The extern modifier is most commonly used when there are two or more files sharing the same global variables or functions as explained below.

First File: main.c

```
#include <stdio.h>
int count ;
extern void write_extern();
 main() { count
 = 5;
 write_extern();
}
```

Second File: support.c

```
#include <stdio.h>
 extern int count;
 void write_extern(void) {
 printf("count is %d\n", count);
}
```

Here, extern is being used to declare count in the second file, where as it has its definition in the first file, main.c. Now, compile these two files as follows:

```
$gcc main.c support.c
```

It will produce the executable program **a.out**. When this program is executed, it produces the following result.

```
count is 5
```

1.5. Constants in C Programming

- C Constants are also like normal variables. But, only difference is, their values can not be modified by the program once they are defined.
- Constants refer to fixed values. They are also called as literals
- Constants may be belonging to any of the data type.

Syntax:

const data_type variable_name; (or) *const data_type *variable_name;*

Types of C Constant

There are 6 constants in C programming. There are

- Integer constants
- Real or Floating point constants
- Octal & Hexadecimal constants
- Character constants
- String constants
- Backslash character constants

Constant type	data type (Example)
Integer constants	int (53, 762, -478 etc) unsigned int (5000u, 1000U etc) long int, long long int (483,647 2,147,483,680)
Real or Floating point constants	float (10.456789) doule (600.123456789)
Octal constant	int (Example: 013 /*starts with 0 */)
Hexadecimal constant	int (Example: 0x90 /*starts with 0x*/)
character constants	char (Example: _A', _B', _C')
string constants	char (Example: —ABCD‖, —Hai‖)

1.5.1. Rules for constructing C constant

Integer Constants in C

- An integer constant must have at least one digit.

- It must not have a decimal point.
- It can either be positive or negative.
- No commas or blanks are allowed within an integer constant.
- If no sign precedes an integer constant, it is assumed to be positive.
- The allowable range for integer constants is -32768 to 32767.

Real Constants in C

- A real constant must have at least one digit
- It must have a decimal point
- It could be either positive or negative
- If no sign precedes an integer constant, it is assumed to be positive.
- No commas or blanks are allowed within a real constant.

Character and String Constants in C

- A character constant is a single alphabet, a single digit or a single special symbol enclosed within single quotes.
- The maximum length of a character constant is 1 character.
- String constants areenclosed within double quotes.

Backslash Character Constants in C

- There are some characters which have special meaning in C language.
- They should be preceded by backslash symbol to make use of special function of them.
- Given below is the list of special characters and their purpose.

Backslash_character	Meaning
\b	Backspace
\f	Form feed
\n	New line
\r	Carriage return
\t	Horizontal tab
\"	Double quote

\‘	Single quote
\\	Backslash
\v	Vertical tab
\a	Alert or bell
\?	Question mark
\N	Octal constant (N is an octal constant)
\XN	Hexadecimal constant (N – hex.dcml cnst)

1.5.2. How to use constants in a C program?

We can define constants in a C program in the following ways.

- By —const‖ keyword
- By —#define‖ preprocessor directive

Please note that when you try to change constant values after defining in C program, it will through error.

Example program using const keyword in C

```
#include <stdio.h>
void main()
{
const int height = 100; /*int constant*/
const float number = 3.14; /*Real constant*/
const char letter = 'A'; /*char constant*/
const char letter_sequence[10] = "ABC"; /*string constant*/
const char backslash_char = '\?'; /*special char cnst*/
printf("value of height :%d \n", height );
printf("value of number : %f \n", number );
printf("value of letter : %c \n", letter );
printf("value of letter_sequence : %s \n", letter_sequence);
printf("value of backslash_char : %c \n", backslash_char);
}
```

Output:

```
value of height : 100
value of number : 3.140000
value of letter : A
value of letter_sequence : ABC
value of backslash_char : ?
```

Example program using #define preprocessor directive in C

```
#include <stdio.h>
#define height 100
#define number 3.14
#define letter 'A'
#define letter_sequence "ABC"
#define backslash_char '\?'
void main()
{
printf("value of height : %d \n", height );
printf("value of number : %f \n", number );
printf("value of letter : %c \n", letter );
printf("value of letter_sequence : %s \n",letter_sequence);
printf("value of backslash_char : %c \n",backslash_char);
}
```

Output

```
value of height : 100
    value of number : 3.140000
value of letter : A
value of letter_sequence : ABC
value of backslash_char : ?
```

1.6. Enumeration Constants

- An **enum** is a keyword, it is an user defined data type. All properties of integer are applied on Enumeration data type so size of the enumerator data type is 2 byte. It work like the Integer.
- It is used for creating an user defined data type of integer. Using enum we can create sequence of integer constant value.

Syntax

```
enum tagname {value1, value2, value3,....};
```

- In above syntax enum is a keyword. It is a user defiend data type.
- In above syntax tagname is our own variable. tagname is any variable name.
- value1, value2, value3,.... are create set of enum values.

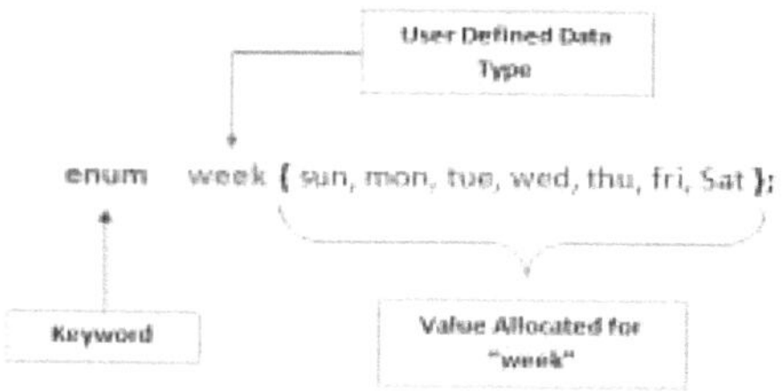

Figure 2: Enumeration data type

Example of Enumeration in C

```
enum week {sun, mon, tue, wed, thu, fri, sat};
enum week today;
```

In above code first line is create user defined data type called week.

- week variable have 7 value which is inside { } braces.
- today variable is declare as week type which can be initialize any data or value among 7 (sun, mon,....).

Example of Enumeration in C

```
#include<stdio.h>
#include<conio.h>

enum ABC {x,y,z};
void main()
{
int a;
clrscr();
a=x+y+z;          //0+1+2
printf("Sum: %d",a);
getch();
}
```

Output

```
Sum: 3
```

Example of Enumeration in C

```
#include<stdio.h>
#include<conio.h>

enum week {sun, mon, tue, wed, thu, fri, sat};
```

```
void main()
{
 enum week today;
 today=tue;
 printf("%d day",today+1);
 getch();
}
```

Output

```
3 day
```

Here "enum week" is user defined data type and today is integer type variable which initialize Tuesday. In above code we add "today+1" because enum is start from 0, so to get proper answer we add "1". Other wise it give "2nd day".

Example of Enumeration in C

```
#include<stdio.h>
#include<conio.h>

enum week {sun, mon, tue, wed, thu, fri, sat};
void main()
{
for(i=sun; i<=sat; i++)
{
 printf("%d ",i);
}
getch();
}
```

Output

```
1 2 3 4 5 6 7
```

In above code replace sun, mon, tue,.... with Equivalent numeric value 0, 1, 2,...

1.7. Keywords in C Programming

There are 32 keywords in C programming. These are:

auto	**break**	**case**	**char**	**const**	**continue**
default	**do**	**double**	**else**	**enum**	**extern**
float	**for**	**goto**	**if**	**Int**	**long**
register	**return**	**short**	**signed**	**sizeof**	**static**
struct	**switch**	**typedef**	**union**	**unsigned**	**void**
volatile	**while**				

1.7.1. Description of all keywords in C

auto

The auto keyword declares automatic variables. For example:

```
auto int var1;
```

- This statement suggests that var1 is a variable of storage class auto and type int.
- Variables declared within function bodies are automatic by default. They are recreated each time a function is executed.
- Since, automatic variables are local to a function, they are also called local variables

break and continue

- The break statement makes program jump out of the innermost enclosing loop (while, do, for or switch statements) explicitly.
- The continue statement skips the certain statements inside the loop.

```
for (i=1;i<=10;++i)
{
 if (i==3)
 continue;
 if (i==7)
 break;
 printf(" %d ",i);
}
```

Output

```
1 2 4 5 6
```

When i is equal to 3, continue statement comes into effect and skips 3. When i is equal to 7, break statement comes into effect and terminates the for loop.

switch, case and default

The switch and case statement is used when a block of statements has to be executed among many blocks.

For example

```
switch(expression)
{
case '1':
//some statements to execute when 1
break;
case '5':
//some statements to execute when 5
break;
default:
//some statements to execute when default;
}
```

char

The char keyword declares a character variable. For example:

```
char alphabet;
```

Here, alphabet is a character type variable.

const

An identifier can be declared constant by using const keyword.

```
const int a = 5;
```

do...while

```
int i;
do
{
 print("%d ",i);
 i++;
}
while (i<10)
```

double and float

Keywords double and float are used for declaring floating type variables. For example:

```
float number;
double longNumber;
```

Here, number is single precision floating type variable whereas, longNumber is a double precision floating type variable.

if and else

In C programming, if and else are used to make decisions.

```
if (i == 1)
printf("i is 1.")
else
 prinf("i is not 1.")
```

If value of i is other than 1, output will be:

```
i is not 1
```

enum

Enumeration types are declared in C programming using keyword enum. For example:

```
enum suit
{ hearts;
spades;
clubs;
diamonds;
};
```

Here, a enumerated variable suit is created having tags: hearts, spades, clubs and diamonds.

extern

The extern keyword declares that a variable or a function has external linkage outside of the file it is declared.

for

There are three types of loops in C programming. The for loop is written in C programming using keyword for. For example:

```
for (i=0; i< 9;++i)
{
printf("%d ",i);
}
```

Output

```
0 1 2 3 4 5 6 7 8
```

goto

The goto keyword is used for unconditional jump to a labeled statement inside a function. For example:

```
for(i=1; i<5; ++i)
{
if (i==10)
goto error;
}
printf("i is not 10");
error:
printf("Error, count cannot be 10.");
```

Output

```
Error, count cannot be 10.
```

int

The int keyword declares integer type variable. For example:

```
int count;
```

Here, count is a integer variable.

short, long, signed and unsigned

The short, long, signed and unsigned keywodrs are type modifiers that alters the meaning of a base data type to yield a new type.

```
short int smallInteger;
long int bigInteger;
signed int normalInteger;
unsigned int positiveInteger;
```

Range of int type data types	
Data types	**Range**
short int	-32768 to 32767
long int	-2147483648 to 214743648
signed int	-32768 to 32767
unsigned int	0 to 65535

return

The return keyword terminates the function and returns the value.

```
int func()
{
int b = 5;
```

```
return b;
}
```

This function func() returns 5 to the calling function.

sizeof

The sizeof keyword evaluates the size of data (a variable or a constant).

```
#include <stdio.h>
int main()
{
printf(" %u bytes.",sizeof(char));
}
```

Output

```
1 bytes.
```

register

The register keyword creates register variables which are much faster than normal variables.

```
register int var1;
```

static

The static keyword creates static variable. The value of the static variables persists until the end of the program. For example:

```
static int var;
```

struct

The struct keyword is used for declaring a structure. A structure can hold variables of different types under a single name.

```
struct student{
char name[80];
float marks;
 int age;
}s1, s2;
```

typedef

The typedef keyword is used to explicitly associate a type with an identifier.

```
typedef float kg;
kg bear, tiger;
```

union

A Union is used for grouping different types of variable under a single name.

```
union student
{
char name[80];
float marks;
int age;
}
```

void

The void keyword indicates that a function doesn't return any value.

```
void testFunction(int a)
{
.....
}
```

Here, function testFunction( ) cannot return a value because the return type is void.

volatile

The volatile keyword is used for creating volatile objects. A volatile object can be modified in an unspecified way by the hardware.

```
const volatile number
```

Here, *number* is a volatile object.

Since, number is a constant variable, the program cannot change it. However, hardware can change it since it is a volatile object.

1.8. Operators in C

- C programming has various operators to perform tasks including arithmetic, conditional and bitwise operations. You will learn about various C operators and how to use them in this tutorial.
- An operator is a symbol which operates on a value or a variable. For example: + is an operator to perform addition.

C programming has wide range of operators to perform various operations. For better understanding of operators, these operators can be classified as:

1.8.1. Operators in C programming

- Arithmetic Operators
- Increment and Decrement Operators
- Assignment Operators
- Relational Operators
- Logical Operators
- Conditional Operators
- Bitwise Operators
- Special Operators

1.8.2. C arithmetic operators

An arithmetic operator performs mathematical operations such as addition, subtraction and multiplication on numerical values (constants and variables).

Operator	Meaning of Operator
+	addition or unary plus
-	subtraction or unary minus
*	multiplication
/	division
%	remainder after division(modulo division)

Example of Arithmetic Operators

```
// C Program to demonstrate the working of arithmetic operators
#include <stdio.h>
int main()
{
int a = 9,b = 4, c;

c = a+b;
printf("a+b = %d \n",c);

c = a-b;
printf("a-b = %d \n",c);

c = a*b;
printf("a*b = %d \n",c);
```

```
c=a/b;
printf("a/b = %d \n",c);

c=a%b;
printf("Remainder when a divided by b = %d \n",c);

return 0;
}
```

Output

```
a+b = 13
a-b = 5
a*b = 36
a/b = 2
Remainder when a divided by b=1
```

- The operators +, - and * computes addition, subtraction and multiplication respectively as you might have expected.
- In normal calculation, 9/4 = 2.25. However, the output is 2 in the program.
- It is because both variables a and b are integers. Hence, the output is also an integer. The compiler neglects the term after decimal point and shows answer 2 instead of 2.25.
- The modulo operator % computes the remainder. When a = 9 is divided by b = 4, the remainder is 1. The % operator can only be used with integers.
- Suppose a = 5.0, b = 2.0, c = 5 and d = 2. Then in C programming,

```
a/b = 2.5// Because both operands are floating-point variables
a/d = 2.5// Because one operand is floating-point variable
c/b = 2.5// Because one operand is floating-point variable
c/d = 2 // Because both operands are integers
```

1.8.3. Increment and decrement operators

- C programming has two operators increment ++ and decrement -- to change the value of an operand (constant or variable) by 1.

- Increment ++ increases the value by 1 whereas decrement -- decreases the value by 1.
- These two operators are unary operators, meaning they only operate on a single operand.

Example of Increment and Decrement Operators

```
// C Program to demonstrate the working of increment and decrement operators
#include <stdio.h>
int main()
{
int a = 10, b = 100;
float c = 10.5, d = 100.5;
printf("++a = %d \n", ++a);
printf("--b = %d \n", --b);
printf("++c = %f \n", ++c);
printf("--d = %f \n", --d);
return 0;
}
```

Output

```
++a = 11
--b = 99
++c = 11.500000
++d = 99.500000
```

Here, the operators ++ and -- are used as prefix. These two operators can also be used as postfix like a++ and a--.

1.8.4. C assignment operators

An assignment operator is used for assigning a value to a variable. The most common assignment operator is =

Operator	Example	Same as
=	a = b	a = b
+=	a += b	a = a+b
-=	a -= b	a = a-b
*=	a *= b	a = a*b
/=	a /= b	a = a/b
%=	a %= b	a = a%b

Example of Assignment Operators

```
// C Program to demonstrate the working of assignment operators
#include <stdio.h>
int main()
{
int a = 5, c;

c = a;
printf("c = %d \n", c);

c += a; // c = c+a
printf("c = %d \n", c);

c -= a; // c = c-a
printf("c = %d \n", c);

c *= a; // c = c*a
printf("c = %d \n", c);

c /= a; // c = c/a
printf("c = %d \n", c);

c %= a; // c = c%a
printf("c = %d \n", c);

return 0;
}
```

Output

```
c = 5
c = 10
c = 5
c = 25
c = 5
c = 0
```

1.8.5. C relational operators

- A relational operator checks the relationship between two operands. If the relation is true, it returns 1; if the relation is false, it returns value 0.
- Relational operators are used in decision making and loops.

Operator	Meaning of Operator	Example
==	Equal to	5 == 3 returns 0
>	Greater than	5 > 3 returns 1
<	Less than	5 < 3 returns 0
!=	Not equal to	5 != 3 returns 1
>=	Greater than or equal to	5 >= 3 returns 1
<=	Less than or equal to	5 <= 3 return 0

Example of Relational Operators

```
// C Program to demonstrate the working of arithmetic operators
#include <stdio.h>
int main()
{
int a = 5, b = 5, c = 10;

printf("%d == %d = %d \n", a, b, a == b); // true
printf("%d == %d = %d \n", a, c, a == c); // false

printf("%d > %d = %d \n", a, b, a > b); //false
printf("%d > %d = %d \n", a, c, a > c); //false

printf("%d < %d = %d \n", a, b, a < b); //false
printf("%d < %d = %d \n", a, c, a < c); //true

printf("%d != %d = %d \n", a, b, a != b); //false
printf("%d != %d = %d \n", a, c, a != c); //true

printf("%d >= %d = %d \n", a, b, a >= b); //true
printf("%d >= %d = %d \n", a, c, a >= c); //false

printf("%d <= %d = %d \n", a, b, a <= b); //true
printf("%d <= %d = %d \n", a, c, a <= c); //true
return 0;
}
```

Output

```
5 == 5 = 1
5 == 10 = 0
5 > 5 = 0
5 > 10 = 0
```

```
5 < 5 = 0
5 < 10 = 1
5 != 5 = 0
5 != 10 = 1
5 >= 5 = 1
5 >= 10 = 0
5 <= 5 = 1
5 <= 10 = 1
```

1.8.6. C logical operators

- An expression containing logical operator returns either 0 or 1 depending upon whether expression results true or false.
- Logical operators are commonly used in decision making in C programming.

Operator	Meaning of Operator	Example
&&	Logial AND. True only if all operands are true	If c = 5 and d = 2 then, expression ((c == 5) && (d > 5)) equals to 0.
||	Logical OR. True only if either one operand is true	If c = 5 and d = 2 then, expression ((c == 5) || (d > 5)) equals to 1.
!	Logical NOT. True only if the operand is 0	If c = 5 then, expression ! (c == 5) equals to 0.

Example of Logical Operators

```
// C Program to demonstrate the working of logical operators
#include <stdio.h>
int main()
{
int a = 5, b = 5, c = 10, result;

result = (a == b) && (c > b);
printf("(a == b) && (c > b) equals to %d \n", result);

result = (a == b) && (c < b);
printf("(a == b) && (c < b) equals to %d \n", result);

result = (a == b) || (c < b);
printf("(a == b) || (c < b) equals to %d \n", result);

result = (a != b) || (c < b);
printf("(a != b) || (c < b) equals to %d \n", result);
```

```
result = !(a != b);
printf("!(a == b) equals to %d \n", result);

result = !(a == b);
printf("!(a == b) equals to %d \n", result);

return 0;
}
```

Output

```
(a == b) && (c > b) equals to 1
(a == b) && (c < b) equals to 0
(a == b) || (c < b) equals to 1
(a != b) || (c < b) equals to 0
!(a != b) equals to 1
!(a == b) equals to 0
```

1.8.7. Explanation of logical operator program

- (a == b) && (c > 5) evaluates to 1 because both operands (a == b) and (c > b) is 1 (true).
- (a == b) && (c < b) evaluates to 0 because operand (c < b)is 0 (false).
- (a == b) | | (c < b) evaluates to 1 because (a = b) is 1 (true).
- (a != b) | | (c < b) evaluates to 0 because both operand (a != b) and (c < b) are 0 (false).
- !(a != b) evaluates to 1 because operand (a != b) is 0 (false). Hence, !(a != b) is 1 (true).
- !(a == b) evaluates to 0 because (a == b) is 1 (true). Hence, !(a == b) is 0 (false).

1.8.8. Bitwise operators

- During computation, mathematical operations like: addition, subtraction, addition and division are converted to bit-level which makes processing faster and saves power.
- Bitwise operators are used in C programming to perform bit-level operations.

Operators	Meaning of operators
&	Bitwise AND
|	Bitwise OR
^	Bitwise exclusive OR

~	Bitwise complement
<<	Shift left
>>	Shift right

1.8.9. Bitwise AND *operator* &

The output of bitwise AND is 1 if the corresponding bits of two operands is 1. If either bit of an operand is 0, the result of corresponding bit is evaluated to 0.

Let us suppose the bitwise AND operation of two integers 12 and 25.

12 = 00001100 (In Binary)

25 = 00011001 (In Binary)

Bit Operation of 12 and 25

00001100

& 00011001

____

00001000= 8 (In decimal)

Example of Bitwise AND

```
#include <stdio.h>
int main()
{
int a = 12, b = 25;
printf("Output = %d", a&b);
return 0;
}
```

Output

```
Output = 8
```

Bitwise OR operator |

The output of bitwise OR is 1 if at least one corresponding bit of two operands is 1. In C Programming, bitwise OR operator is denoted by |.

12 = 00001100 (In Binary)

25 = 00011001 (In Binary)

Bitwise OR Operation of 12 and 25

00001100
| 00011001

____

00011101= 29 (In decimal)

Example of Bitwise OR

```
#include <stdio.h>
int main()
{
int a = 12, b = 25;
printf("Output = %d", a | b);
return 0;
}
```

Output

```
Output = 29
```

1.8.10. Bitwise XOR (exclusive OR) operator ^

The result of bitwise XOR operator is 1 if the corresponding bits of two operands are opposite. It is denoted by ^.

12 = 00001100 (In Binary)
25 = 00011001 (In Binary)

Bitwise XOR Operation of 12 and 25
00001100
| 00011001

____

00010101= 21 (In decimal)

Example of Bitwise XOR

```
#include <stdio.h>
int main()
{
int a = 12, b = 25;
printf("Output = %d", a^b);
return 0;
}
```

Output

Output = 21

1.8.11. Bitwise complement operator ~

Bitwise compliment operator is an unary operator (works on only one operand). It changes 1 to 0 and 0 to 1. It is denoted by ~.

35 = 00100011 (In Binary)

Bitwise complement Operation of 35

~ 00100011

____

11011100= 220 (In decimal)

Twist in bitwise complement operator in C Programming

- The bitwise complement of 35 (~35) is -36 instead of 220, but why?
- For any integer n, bitwise complement of n will be -(n+1). To understand this, you should have the knowledge of 2's complement.

2's Complement

Two's complement is an operation on binary numbers. The 2's complement of a number is equal to the complement of that number plus 1.

For example:

Decimal	Binary	2's complement
0	00000000	-(11111111+1) = -00000000 = -0(decimal)
1	00000001	-(11111110+1) = -11111111 = -256(decimal)
12	00001100	-(11110011+1) = -11110100 = -244(decimal)
220	11011100	-(00100011+1) = -00100100 = -36(decimal)

Note: Overflow is ignored while computing 2's complement.

- The bitwise complement of 35 is 220 (in decimal). The 2's complement of 220 is -36. Hence, the output is -36 instead of 220.
- Bitwise complement of any number N is -(N+1). Here's how:
- bitwise complement of N = ~N (represented in 2's complement form)
- 2'complement of ~N= -(~(~N)+1) = -(N+1)

Example of Bitwise complement

```
#include <stdio.h>
int main()
{
printf("complement = %d\n",~35);
printf("complement = %d\n",~-12);
return 0;
}
```

Output

```
complement = -36
Output = 11
```

1.8.12. Shift operators in C programming

There are two shift operators in C programming:

- Right shift operator
- Left shift operator.

Right Shift Operator

Right shift operator shifts all bits towards right by certain number of specified bits. It is denoted by >>.

212 = 11010100 (In binary)
212>>2 = 00110101 (In binary) [Right shift by two bits]
212>>7 = 00000001 (In binary)
212>>8 = 00000000
212>>0 = 11010100 (No Shift)

Left Shift Operator

Left shift operator shifts all bits towards left by certain number of specified bits. It is denoted by <<.

212 = 11010100 (In binary)
212<<1 = 110101000 (In binary) [Left shift by one bit]
212<<0 =11010100 (Shift by 0)
212<<4 = 110101000000 (In binary) =3392(In decimal)

Example of Shift Operators

```
#include <stdio.h>
int main()
{
int num=212, i;
for (i=0; i<=2; ++i)
printf("Right shift by %d: %d\n", i, num>>i);

 printf("\n");

 for (i=0; i<=2; ++i)
printf("Left shift by %d: %d\n", i, num<<i);

 return 0;
}
```

Output:

```
Right Shift by 0: 212
Right Shift by 1: 106
Right Shift by 2: 53

Left Shift by 0: 212
Left Shift by 1: 424
Left Shift by 2: 848
```

1.8.13. Other operators

Comma Operator

Comma operators are used to link related expressions together.

For example: int a, c = 5, d;

The sizeof operator

The sizeof is an unary operator which returns the size of data (constant, variables, array, structure etc).

Example of sizeof Operator

```
#include <stdio.h>
int main()
{
int a, e[10];
float b;
double c;
char d;
printf("Size of int=%lu bytes\n",sizeof(a));
printf("Size of float=%lu bytes\n",sizeof(b));
printf("Size of double=%lu bytes\n",sizeof(c));
printf("Size of char=%lu byte\n",sizeof(d));
printf("Size of integer type array having 10 elements = %lu bytes\n",
sizeof(e));
return 0;
}
```

Output

```
Size of int = 4 bytes
Size of float = 4 bytes
Size of double = 8 bytes
Size of char = 1 byte
Size of integer type array having 10 elements = 40 bytes
```

C Ternary Operator (?:)

A conditional operator is a ternary operator, that is, it works on 3 operands.

Conditional Operator Syntax

```
conditionalExpression ? expression1 : expression2
```

The conditional operator works as follows:

- The first expression conditionalExpression is evaluated first. This expression evaluates to 1 if it's true and evaluates to 0 if it's false.
- If conditionalExpression is true, expression1 is evaluated.
- If conditionalExpression is false, expression2 is evaluated.

Example of C conditional Operator

```
#include <stdio.h>
int main(){
 char February;
 int days;
 printf("If this year is leap year, enter 1. If not enter any integer: ");
 scanf("%c",&February);

 // If test condition (February == '1') is true, days equal to 29.
 // If test condition (February =='1') is false, days equal to 28.
 days = (February == '1') ? 29 : 28;

 printf("Number of days in February = %d",days);
 return 0;
}
```

Output

```
If this year is leap year, enter 1. If not enter any integer: 1
Number of days in February = 29
```

1.9. Operators Precedence in C

- Operator precedence determines the grouping of terms in an expression and decides how an expression is evaluated. Certain operators have higher precedence than others; for example, the multiplication operator has a higher precedence than the addition operator.
- For example, x = 7 + 3 * 2; here, x is assigned 13, not 20 because operator * has a higher precedence than +, so it first gets multiplied with 3*2 and then adds into 7.
- Here, operators with the highest precedence appear at the top of the table, those with the lowest appear at the bottom. Within an expression, higher precedence operators will be evaluated first.

Category	Operator	Associativity
Postfix	() [] -> . ++ - -	Left to right
Unary	+ - ! ~ ++ - - (type)* & sizeof	Right to left
Multiplicative	* / %	Left to right
Additive	+ -	Left to right
Shift	<< >>	Left to right

Relational	< <= > >=	Left to right
Equality	== !=	Left to right
Bitwise AND	&	Left to right
Bitwise XOR	^	Left to right
Bitwise OR	|	Left to right
Logical AND	&&	Left to right
Logical OR	||	Left to right
Conditional	?:	Right to left
Assignment	= += -= *= /= %=>>= <<= &= ^= |=	Right to left
Comma	,	Left to right

Example

Try the following example to understand operator precedence in C .

```
#include <stdio.h>
main() {
 int a = 20;
 int b = 10;
 int c = 15;
 int d = 5;
 int e;

e = (a + b) * c / d;// ( 30 * 15 ) / 5
 printf(" Value of (a + b) * c / d is : %d\n",e );

 e = ((a + b) * c) / d;// (30 * 15 ) / 5
 printf(" Value of ((a + b) * c) / d is: %d\n" ,e );

 e = (a + b) * (c / d); // (30) * (15/5)
 printf(" Value of (a + b) * (c / d) is: %d\n",e );

 e = a + (b * c) / d; //20 + (150/5)
 printf(" Value of a + (b * c) / d is: %d\n" ,e );

 return 0;
}
```

Output:

```
Value of (a + b) * c / d is : 90
Value of ((a + b) * c) / d is: 90
Value of (a + b) * (c / d) is: 90
Value of a + (b * c) / d is: 50
```

Summary of operator precedence

- Comma Operator Has Lowest Precedence .
- Unary Operators are Operators having Highest Precedence.
- Sizeof is Operator not Function .
- Operators sharing Common Block in the Above Table have Equal Priority or Precedence.
- While Solving Expression , Equal Priority Operators are handled on the basis of FIFO [First in First Out] i.e Operator Coming First is handled First.

Important definitions

Unary Operator

- A unary operation is an operation with only one operand, i.e. an operation with a single input.
- A unary operator is one which has only one operand. e.g. post / pre increment operator

Binary Operator

- A Binary operator is one which has two operand. e.g. plus , Minus .

Associativity

Their associativity indicates in what order operators of equal precedence in an expression are applied

Precedence

- Priority of Operator

Examples : operator precedence & associativity

We have listed out some of the examples based on operator precedence & associativity. Examples will give you clear picture how to use operator precedence & associativity table chart while solving the expression

Example of Simple use of precedence chart

```
#include<stdio.h>
int main() {
int num1 = 10, num2 = 20;
int result;

result = num1 * 2 + num2;
printf(" \nResult is : %d", result);

return (0);
}
```

Consider the simple expression used in above program and refer operator precedence & associativity table chart

```
result = num1 * 2 + num2;
```

In this case multiplication operator will have higher priority than addition and assignment operator so multiplication will be evaluated firstly.

```
result = num1 * 2 + num2;
result = 10 * 2 + 20;
result = 20 + 20;
result = 40;
```

Example of Use of associativity

In above example none of the operator has equal priority. In the below example some operators are having same priority.

```
result = num1 * 2 + num2 * 2 ;
```

In the above expression we have overall 3 operators i.e. 2 multiplication, 1 addition and 1 assignment operator.

Now refer operator precedence & associativity table (block 3) which clearly tells that associativity is from left to right so we will give priority to multiplication operator on the left side

```
result = num1 * 2 + num2 * 2 ;
result = 10 * 2 + 20 * 2 ;
result = 20 + 20 * 2 ;
result = 20 + 40 ;
result = 60 ;
```

1.10. C Programming Expression

- In programming, an expression is any legal combination of symbols that represents a value.
- C Programming Provides its own rules of Expression, whether it is legal expression or illegal expression. For example, in the C language x+5 is a legal expression.
- Every expression consists of at least one operand and can have one or more operators.
- Operands are values and Operators are symbols that represent particular actions.

Valid C Programming Expression

C Programming code gets compiled firstly before execution. In the different phases of compiler, c programming expression is checked for its validity.

Expressions	Validity
a + b	Expression is valid since it contain + operator which is binary operator
+ + a + b	Invalid Expression

Types of Expression

In Programming, different verities of expressions are given to the compiler. Expressions can be classified on the basis of Position of Operators in an expression.

Type	Explanation	Example
Infix	Expression in which Operator is in between Operands	a + b
Prefix	Expression in which Operator is written before Operands	+ a b
Postfix	Expression in which Operator is written after Operands	a b +

These expressions are solved using the stack.

Examples of Expression:

- Now we will be looking into some of the C Programming Expressions, Expression can be created by combining the operators and operands
- Each of the expression results into the some resultant output value. Consider few expressions in below table

Expression Examples, Explanation

- n1 + n2,This is an expression which is going to add two numbers and we can assign the result of addition to another variable.
- x = y,This is an expression which assigns the value of right hand side operand to left side variable
- v = u + a * t,We are multiplying two numbers and result is added to ‗u‘ and total result is assigned to v
- x <= y,This expression will return Boolean value because comparison operator will give us output either true or false ++j,This is expression having pre increment operator\, it is used to increment the value of j before using it in expression [/table].

1.11. Input/Output Statements in C Programming

- There are some library functions which are available for transferring the information between the computer and the standard input and output devices.
- These functions are related to the symbolic constants and are available in the header file.

  Some of the input and output functions are as follows:

Printf()

This function is used for displaying the output on the screen i.e the data is moved from the computer memory to the output device.

Syntax:

printf("format string", arg1, arg2, ...);

In the above syntax, 'format string' will contain the information that is formatted. They are the general characters which will be displayed as they are.
arg1, arg2 are the output data items.
Example: Demonstrating the printf function
printf(—Enter a value:‖);

- Printf() will generally examine from left to right of the string.
- The characters are displayed on the screen in the manner they are encountered until it comes across % or \.
- Once it comes across the conversion specifiers it will take the first argument and print it in the format given.

Scanf()
scanf is used when we enter data by using an input device.

Syntax
scanf ("format string", &arg1, &arg2, ...);

- The number of items which are successful are returned.
- Format string consists of the conversion specifier. Arguments can be variables or array name and represent the address of the variable. Each variable must be preceded by an ampersand (&). Array names should never begin with an ampersand.

Example: Demonstrating scanf()
int avg;
float per;
char grade;
scanf("%d %f %c",&avg, &per, &grade):

- scanf works totally opposite to printf. The input is read, interpret using the conversion specifier and stores it in the given variable.
- The conversion specifier for scanf is the same as printf.
- scanf reads the characters from the input as long as the characters match or it will terminate. The order of the characters that are entered are not important.
- It requires an enter key in order to accept an input.

getch()

This function is used to input a single character. The character is read instantly and it does not require an enter key to be pressed. The character type is returned but it does not echo on the screen.

Syntax

int getch(void);

ch=getch();

where,

ch - assigned the character that is returned by getch.

A simple C program to illustrate the use of getch() function:

```
/*Program to explain the use of getch() function*/
#include
main()
{
char n;
puts("Enter the Char");
n = getch();
puts("Char is :");
putchar(n);
getch();
}
```

Output

```
Enter the Char
Char is L
```

Putch()

This function is a counterpart of getch. Which means that it will display a single character on the screen. The character that is displayed is returned.

Syntax

int putch(int);

putch(ch);

where,

ch - the character that is to be printed.

getche()

This function is used to input a single character. The main difference between getch and getche is that getche displays the (echoes) the character that we type on the screen.

Syntax

int getch(void);
ch=getche();

A simple C program to illustrate the use of getch() function:

```
/*Program to explain the use of getch() function*/
#include
main()
{
char n;
puts("Enter the Char");
n = getche();
puts("Char is :");
putchar(n);
getche();
}
```

Output

```
Enter the Char
Char is L
```

getchar()

This function is used to input a single character. The enter key is pressed which is followed by the character that is typed. The character that is entered is echoed.

Syntax

ch=getchar;

A simple C-program to read a single character from the keyboard is as:

```
/*To read a single character from the keyboard using the getchar() function*/
#include
```

```
main()
{
char n;
n = getchar();
}
```

putchar()

This function is the other side of getchar. A single character is displayed on the screen.

Syntax

putchar(ch);

A simple program is written as below, which will read a single character using getchar() function and display inputted data using putchar() function:

```
/*Program illustrate the use of getchar() and putchar() functions*/
#include
main()
{
char n;
n = getchar();
putchar(n);
}
```

gets() and puts()

- They help in transferring the strings between the computer and the standard input-output devices.
- Only single arguments are accepted.
- The arguments must be such that it represents a string.
- It may include white space characters.
- If gets is used enter key has to be pressed for ending the string.
- The gets and puts function are used to offer simple alternatives of scanf and printf for reading and displaying.

Example

```
#include <stdio.h>
void main()
{
```

```
char line[30];
gets (line);
puts (line);
}
```

A simple C program to illustrate the use of gets() function:

```
/*Program to explain the use of gets() function*/
#include
main()
{
char n[20];
gets(n);
}
```

A simple C program to illustrate the use of puts() function:

```
/*Program to illustrate the concept of puts() with gets() functions*/
#include
main()
{
char name[20];
puts("Enter the Name");
gets(name);
puts("Name is :");
puts(name);
}
```

Output

```
Enter the Name
Laura
Name is:
Laura
```

1.12. C Assignment Operator

Assignment Operator in C Programming Language:

- Assignment Operator is Used to assign value to an variable.
- Assignment Operator is denoted by equal to sign
- Assignment Operator is binary operator which operates on two operands.
- Assignment Operator have Two Values – L-Value and R-Value.Operator copies R-Value into L-Value.

- Assignment Operator have lower precedence than all available operators but has higher precedence than comma Operator.

Pictorial Representation

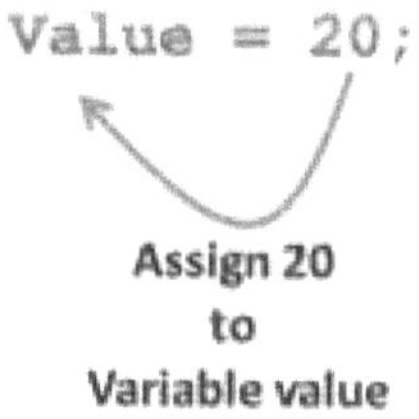

Figure: Assignment Operator

Different Ways of Using Assignment Operator:

Way 1: Assignment Operator used to Assign Value

```
int main() {
 int value;
 value = 55;
 return (0);
}
```

In the above example, we have assigned 55 to a variable _value‘.

Way 2: Assignment Operator used To Type Cast.

```
int value;
value = 55.450;
printf("%d",value);
```

Assignment operator can type cast higher values to lower values. It can also cast lower values to higher values

Way 3: Assignment Operator in If Statement

```
if(value = 10)
printf("True");
else
printf("False");
```

- It would be interesting to understand above program. Above Program will always execute True Condition because Assignment Operator is used inside if statement not comparison operator. Lets see what will happen in above example –

- Constant number 10 will be assigned to variable _value‘. Variable _value‘ will be treated as positive non-zero number, which is considered as true condition.

```
if(value = 0)
printf("True");
else
printf("False");
```

In below case else block will be executed

Shorthand assignment operator

Shorthand assignment operator is used express the syntax that are already available in shorter way

Operator symbol	Name of the operator	Example	Equivalent construct
+=	Addition assignment	x += 4;	x = x + 4;
-=	Subtraction assignment	x -= 4;	x = x – 4;
*=	Multiplication assignment	x *= 4;	x = x * 4;
/=	Division assignment	x /= 4;	x = x / 4;
%=	Remainder assignment	x %= 4;	x = x % 4;

Example : Shorthand assignment operator

```
#include<stdio.h>

int main() {

 int res;

 res = 11;
 res += 4;
 printf("\nResult of Ex1 = %d", res);

 res = 11;
 res -= 4;
```

```
printf("\nResult of Ex2 = %d", res);

res = 11;
res *= 4;
printf("\nResult of Ex3 = %d", res);

res = 11;
res /= 4;
printf("\nResult of Ex4 = %d", res);

res = 11;
res %= 4;
printf("\nResult of Ex5 = %d", res);

return 0;
}
```

Output

```
Result of Ex1 = 15
Result of Ex2 = 7
Result of Ex3 = 44
Result of Ex4 = 2
Result of Ex5 = 3
```

1.13. Decision Making Statement

- Decision making statement is depending on the condition block need to be executed or not which is decided by condition.
- If the condition is "true" statement block will be executed, if condition is "false" then statement block will not be executed.
- In this section we are discuss about if-then (if), if-then-else (if else), and switch statement. In C language there are three types of decision making statement.
- if
- if-else
- switch

if-then Statement

if-then is most basic statement of Decision making statement. It tells to program to execute a certain part of code only if particular condition is true.

Syntax

if(condition)
{
....
....
}

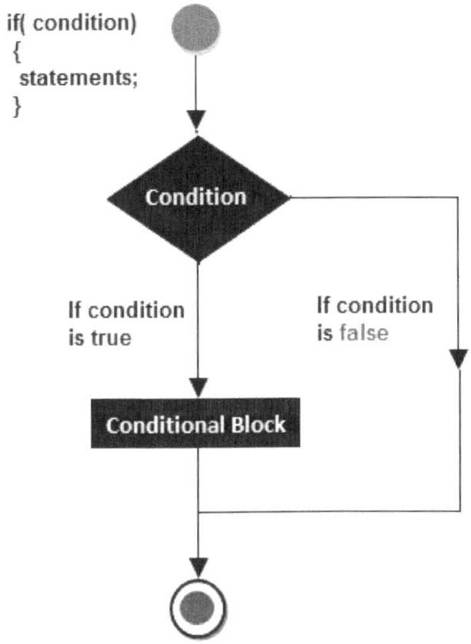

Figure: if statement

- Constructing the body of "if" statement is always optional, Create the body when we are having multiple statements.
- For a single statement, it is not required to specify the body.
- If the body is not specified, then automatically condition part will be terminated with next semicolon (;).

else

- It is a keyword, by using this keyword we can create a alternative block for "if" part. Using else is always optional i.e, it is recommended to use when we are having alternate block of condition.

- In any program among if and else only one block will be executed. When if condition is false then else part will be executed, if part is executed then automatically else part will be ignored.

if-else statement

In general it can be used to execute one block of statement among two blocks, in C language if and else are the keyword in C.

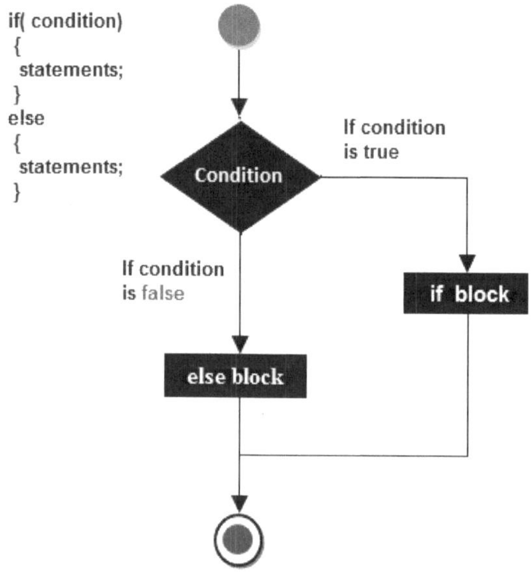

Figure: if-else statement

Syntax

```
if(condition)
{
........
statements
........
}
else
{
........
```

```
statements
........
}
```

In the above syntax whenever condition is true the entire if block statement are executed remaining statement of the program by neglecting else block statement.

If the condition is false else block statement remaining statement of the program are executed by neglecting if block statements.

Example

```
#include<stdio.h>
#include<conio.h>

void main()
{
int time=10;
clrscr();
if(time>12)
{
printf("Good morning");
}
{
printf("Good after noon");
}
getch();
}
```

Output

```
Good morning
```

C Nested if-else Statements

Nested if else statements plays an important role in C programming, it means you can use conditional statements inside another conditional statement.

The basic format of Nested if else statement is:

Syntax

```
if(test_expression one)
{
 if(test_expression two) {
//Statement block Executes when the boolean test expression two is true.
 }
}
else
{
//else statement block
}
```

Example of a C Program to Demonstrate Nested if else Statement

```
#include<stdio.h>

main()
{
int x=20,y=30;

if(x==20)
{
if(y==30)
{
printf("value of x is 20, and value of y is 30.");
}
}
}
```

Output:

```
value of x is 20, and value of y is 30.
```

else-if ladder

The general form of else-if ladder is

```
if(expression1)
{
 statement block1;
}
else if(expression2)
{
 statement block2;
}
```

```
else if(expression3 )
{
 statement block3;
}
else
 default statement;
```

The expression is tested from the top(of the ladder) downwards. As soon as the true condition is found, the statement associated with it is executed.

Example

```
#include <stdio.h>
#include <conio.h>
void main( )
{
 int a;
 printf("enter a number");
 scanf("%d",&a);
 if( a%5==0 && a%8==0)
 {
printf("divisible by both 5 and 8");
 }
 else if( a%8==0 )
 {
printf("divisible by 8");
 }
 else if(a%5==0)
 {
printf("divisible by 5");
 }
 else
 {
printf("divisible by none");
 }
getch();
}
```

C goto Statement

- So far we have discussed about the if statements and how it is used in C to control statement execution based on some decisions or conditions. The flow of execution also depends on other statements which are not based on conditions that can control the flow.

- C supports a special form of statement that is the goto Statement which is used to branch unconditionally within a program from one point to another. Although it is not a good habit to use goto statement in C, but there may be some situations where the use of goto statement might be desirable.
- gotostatement is used by programmers to change the sequence of execution of a C program by shifting the control to a different part of the same program.

The general form of goto statement is:

Syntax:

goto label;

- Label is an identifier required for goto statement in order to place where the branch is to be made.
- A label is a valid variable name which is followed by a colon and is put immediately before the statement where the control needs to be jumped / transferred unconditionally.

Syntax

```
goto label;
        - - -- - -
        - - - - - - - -

label:

statement - X;
/* This the forward jump of goto statement */
or
label:

        - - -- - -
        - - - - - - - -

goto label;

/*This is the backward jump of goto statement */
```

An Example of a C Program to Demonstrate goto Statement

```
#include<stdio.h>

void main()
{
 int age;

 g: //label name
 printf("you are Eligible\n");
 s: //label name
 printf("you are not Eligible");
 printf("Enter you age:");
 scanf("%d", &age);
 if(age>=18)
goto g; //goto label g
 else
goto s; //goto label s
getch();
}
```

1.14. Switch Statement

A switch statement work with byte, short, char and int primitive data type, it also works with enumerated types and string.

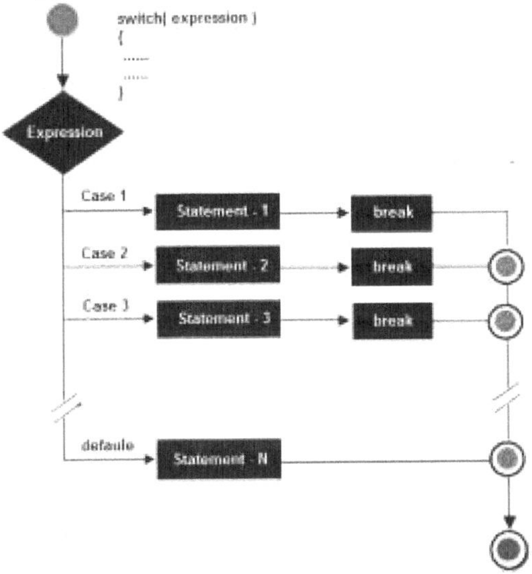

Figure: Switch Statement

Syntax

```
switch(expression/variable)
{
casevalue:
//statements
// any number of case statements
break;//optional
default: //optional
//statements
}
```

Rules for apply switch

- With switch statement use only byte, short, int, char data type.
- You can use any number of case statements within a switch.
- Value for a case must be same as the variable in switch.

Limitations of switch

- Logical operators cannot be used with switch statement. For instance

Example

case k>=20: *//is not allowed*

Switch case variables can have only int and char data type. So float data type is not allowed.

Syntax

```
switch(ch)
{
case1:
statement 1;
break;
case2:
statement 2;
break;
}
```

In this **ch** can be integer or char and cannot be float or any other data type.

Example of Switch case

```
#include<stdio.h>
#include<conio.h>

void main()
{
int ch;
clrscr();
printf("Enter any number (1 to 7)");
scanf("%d",&ch);
switch(ch)
{
case1:
printf("Today is Monday");
break;
case2:
printf("Today is Tuesday");
break;
case3:
printf("Today is Wednesday");
break;
case4:
printf("Today is Thursday");
break;
case5:
printf("Today is Friday");
break;
case6:
printf("Today is Saturday");
break;
case7:
printf("Today is Sunday");
break;
default:
printf("Only enter value 1 to 7");
}
getch();
}
```

Output

```
Enter any number (1 to 7): 5
Today is Friday
```

Note: In switch statement default is optional but when we use this in switch default is executed at last whenever all cases are not satisfied the condition.

1.15. Looping Statement in C

Looping statement are the statements execute one or more statement repeatedly several number of times. In C programming language there are three types of loops; while, for and do-while.

Why use loop?

When you need to execute a block of code several number of times then you need to use looping concept in C language.

Advantage with looping statement

- Reduce length of Code
- Take less memory space.
- Burden on the developer is reducing.
- Time consuming process to execute the program is reduced.

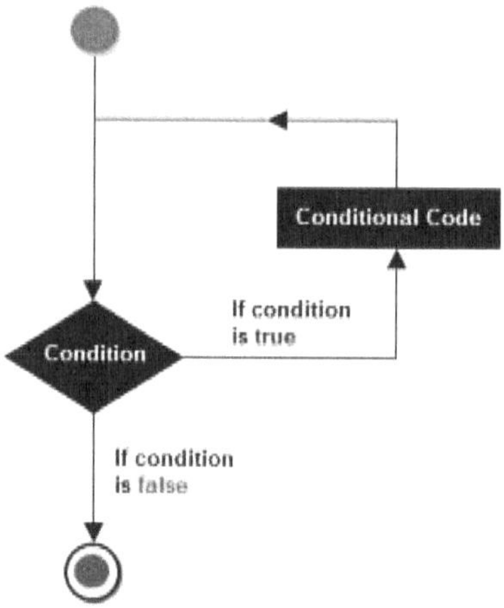

Figure: Looping Statement

Types of Loops

There are three type of Loops available in 'C' programming language.

- while loop
- for loop
- do..while

Difference between conditional and looping statement

Conditional statement executes only once in the program where as looping statements executes repeatedly several number of time.

While loop

- In while loop First check the condition if condition is true then control goes inside the loop body otherwise goes outside the body.
- While loop will be repeats in clock wise direction.

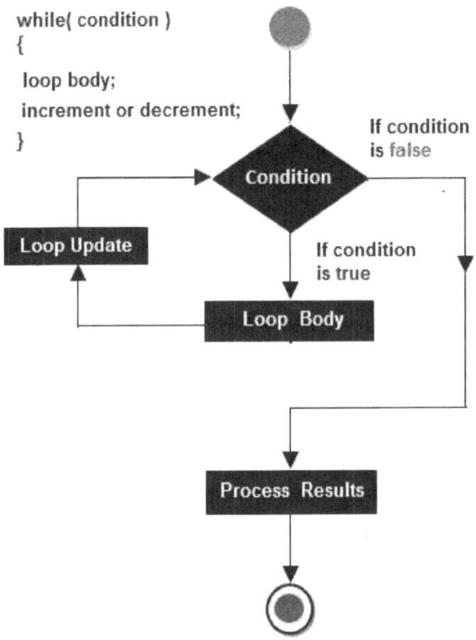

Figure: While Loop

Syntax

```
Assignment;
while(condition)
{
Statements;
......
Increment/decrements (++ or --);
}
```

Note: If while loop condition never false then loop become infinite loop.

Example of while loop

```
#include<stdio.h>
#include<conio.h>
void main()
{
int i;
clrscr();
i=1;
while(i<5)
{
printf("\n%d",i);
i++;
}
getch();
}
```

Output

```
1
2
3
4
```

For loop

for loop is a statement which allows code to be repeatedly executed. For loop contains 3 parts Initialization, Condition and Increment or Decrements

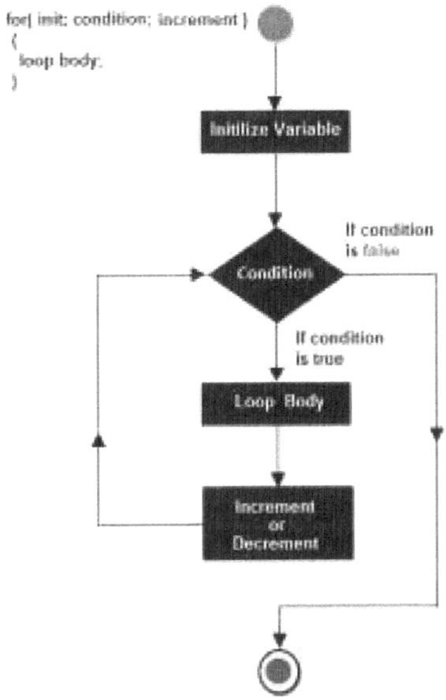

Figure: While Loop

Example of for loop

```
#include<stdio.h>
#include<conio.h>
void main()
{
int i; clrscr();
for(i=1;i<5;i++)
{
printf("\n%d",i);
}
getch();
}
```

Output

```
1
2
3
4
```

do-while

- A **do-while** loop is similar to a while loop, except that a do-while loop is execute at least one time.
- A do while loop is a control flow statement that executes a block of code at least once, and then repeatedly executes the block, or not, depending on a given condition at the end of the block (in while).

Syntax

```
do
{
Statements;
........
Increment/decrement (++ or --)
} while();
```

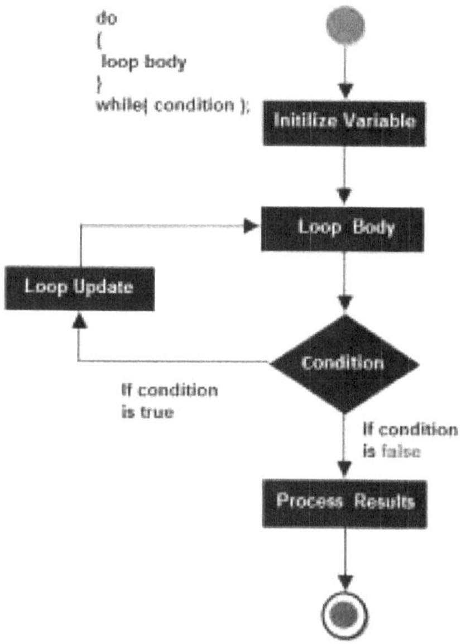

Figure: do...while Loop

When use do...while Loop

When we need to repeat the statement block at least 1 time then we use do-while loop.

Example of do..while loop

```
#include<stdio.h>
#include<conio.h>
void main()
{
int i;
clrscr();
i=1;
do
{
printf("\n%d",i);
i++;
}
while(i<5);
getch();
}
```

Output

```
1
2
3
4
```

Nested loop

- In Nested loop one loop is place within another loop body.
- When we need to repeated loop body itself n number of times use nested loops. Nested loops can be design upto 255 blocks.

1.16. C Preprocessor Directives

The C Preprocessor is not part of the compiler, but is a separate step in the compilation process. In simplistic terms, a C Preprocessor is just a text substitution tool and they instruct compiler to do required pre-processing before actual compilation. We'll refer to the C Preprocessor as the CPP.

All preprocessor commands begin with a pound symbol (#). It must be the first nonblank character, and for readability, a preprocessor directive should begin in first column. Following section lists down all important preprocessor directives:

Directive	Description
#define	Substitutes a preprocessor macro
#include	Inserts a particular header from another file
#undef	Undefines a preprocessor macro
#ifdef	Returns true if this macro is defined
#ifndef	Returns true if this macro is not defined
#if	Tests if a compile time condition is true
#else	The alternative for #if
#elif	#else an #if in one statement
#endif	Ends preprocessor conditional
#error	Prints error message on stderr
#pragma	Issues special commands to the compiler, using a standardized method

Other directives include

- # - Stringization, replaces a macro parameter with a string constant
- ## - Token merge, creates a single token from two adjacent ones

Preprocessors Examples

Analyze the following examples to understand various directives.

```
#define MAX_ARRAY_LENGTH 20
```

This directive tells the CPP to replace instances of MAX_ARRAY_LENGTH with 20. Use #define for constants to increase readability.

```
#include <stdio.h>
#include "myheader.h"
```

These directives tell the CPP to get stdio.h from System Libraries and add the text to the current source file. The next line tells CPP to get myheader.h from the local directory and add the content to the current source file.

```
#undefFILE_SIZE
#define FILE_SIZE 42
```

This tells the CPP to undefine existing FILE_SIZE and define it as 42.

```
#ifndef MESSAGE
 #define MESSAGE "You wish!"
#endif
```

This tells the CPP to define MESSAGE only if MESSAGE isn't already defined.

```
#ifdef DEBUG
 /* Your debugging statements here */
#endif
```

This tells the CPP to do the process the statements enclosed if DEBUG is defined. This is useful if you pass the -DDEBUG flag to gcc compiler at the time of compilation. This will define DEBUG, so you can turn debugging on and off on the fly during compilation.

Predefined Macros

ANSI C defines a number of macros. Although each one is available for your use in programming, the predefined macros should not be directly modified.

Macro	Description
__DATE__	The current date as a character literal in "MMM DD YYYY" format
__TIME__	The current time as a character literal in "HH:MM:SS" format
__FILE__	This contains the current filename as a string literal.
__LINE__	This contains the current line number as a decimal constant.
__STDC__	Defined as 1 when the compiler complies with the ANSI standard.

Example

```
#include <stdio.h>
main()
{
 printf("File :%s\n", __FILE__ );
 printf("Date :%s\n", __DATE__ );
 printf("Time :%s\n", __TIME__ );
 printf("Line :%d\n", __LINE__ );
 printf("ANSI :%d\n", __STDC__ );
}
```

When the above code in a file test.c is compiled and executed, it produces the following result:

```
File :test.c
Date :Jun 2 2012
Time :03:36:24
Line :8
ANSI :1
```

1.16.1. The # AND ## preprocessor operators

The C preprocessor offers following operators to help you in creating macros:

Macro Continuation (\)

A macro usually must be contained on a single line. The macro continuation operator is used to continue a macro that is too long for a single line. For example:

```
#definemessage_for(a, b) \
printf(#a " and " #b ": We love you!\n")
```

The # and ## Preprocessor operators

Stringiz (#)

The string size or number-sign operator ('#'), when used within a macro definition, converts a macro parameter into a string constant. This operator may be used only in a macro that has a specified argument or parameter list. For example:

```
#include <stdio.h>
#definemessage_for(a, b) \
printf(#a " and " #b ": We love you!\n")
int main(void)
{
 message_for(Carole, Debra);
 return 0;
}
```

When the above code is compiled and executed, it produces the following result:

```
Carole and Debra: We love you!
```

Token Pasting (##)

The token-pasting operator (##) within a macro definition combines two arguments. It permits two separate tokens in the macro definition to be joined into a single token. For example:

```
#include <stdio.h>
#define tokenpaster(n) printf ("token" #n " = %d", token##n)
int main(void)
{
 int token34 = 40;
tokenpaster(34);
return 0;
}
```

When the above code is compiled and executed, it produces the following result:

```
token34 = 40
```

How it happened, because this example results in the following actual output from the preprocessor:

```
printf ("token34 = %d", token34);
```

This example shows the concatenation of token##n into token34 and here we have used both stringize and token-pasting.

The defined() Operator

The preprocessor defined operator is used in constant expressions to determine if an identifier is defined using #define. If the specified identifier is defined, the value is true (non-zero). If the symbol is not defined, the value is false (zero). The defined operator is specified as follows:

```
#include <stdio.h>
#if !defined (MESSAGE)
 #define MESSAGE "You wish!"
#endif
int main(void)
{
 printf("Here is the message: %s\n", MESSAGE);
 return 0;
}
```

When the above code is compiled and executed, it produces the following result:

```
Here is the message: You wish!
```

Parameterized Macros:

One of the powerful functions of the CPP is the ability to simulate functions using parameterized macros. For example, we might have some code to square a number as follows:

```
int square(int x)
{
 return x * x;
}
```

We can rewrite above code using a macro as follows:

```
#define square(x) ((x) * (x))
```

Macros with arguments must be defined using the #define directive before they can be used. The argument list is enclosed in parentheses and must immediately follow the macro name. Spaces are not allowed between and macro name and open parentheses.

For example

```
#include <stdio.h>
#define MAX(x,y) ((x) > (y) ? (x) : (y))
int main(void)
{
 printf("Max between 20 and 10 is %d\n", MAX(10, 20));
 return 0;
}
```

When the above code is compiled and executed, it produces the following result:

```
Max between 20 and 10 is 20
```

1.17. Compilation Process in C

Four Stages of Compiling a C Program.

Knowing how compilation works can be very helpful both when writing code and debugging. Compiling a C program is a multi-stage process.

At an overview level, the process can be split into four separate stages:

- Preprocessing
- Compilation
- Assembly
- Linking.

Traditional C compilers orchestrate this process by invoking other programs to handle each stage.

The four stages of compiling the following C program:

```
/*
 * "Hello, World!": A classic.
 */
#include <stdio.h>
int
main(void)
{
        puts("Hello, World!");
        return 0;
}
```

Preprocessing

- The first stage of compilation is called preprocessing. In this stage, lines starting with a # character are interpreted by the preprocessor as preprocessor commands.
- These commands form a simple macro language with its own syntax and semantics.
- This language is used to reduce repetition in source code by providing functionality to inline files, define macros and to conditionally omit code.

- Before interpreting commands, the preprocessor does some initial processing. This includes joining continued lines (lines ending with a \) and stripping comments.

To print the result of the preprocessing stage, pass the -E option to cc:

```
cc -E hello_world.c
```

- Given the —Hello, World!‖ example above, the preprocessor will produce the contents of the stdio.h header file joined with the contents of the hello_world.c file, stripped free from its leading comment:

```
[lines omitted for brevity]
extern int __vsnprintf_chk (char * restrict, size_t,
 int, size_t, const char * restrict, va_list);
# 493 "/usr/include/stdio.h" 2 3 4
# 2 "hello_world.c" 2

int main(void) {
puts("Hello, World!");
return 0;
}
```

Compilation

- The second stage of compilation is confusingly enough called compilation. In this stage, the preprocessed code is translated to assembly instructions specific to the target processor architecture. These form an intermediate human readable language.
- The existence of this step allows for C code to contain inline assembly instructions and for different assemblers to be used.
- Some compilers also supports the use of an integrated assembler, in which the compilation stage generates machine code directly, avoiding the overhead of generating the intermediate assembly instructions and invoking the assembler.

To save the result of the compilation stage, pass the -S option to cc:

```
cc -S hello_world.c
```

- This will create a file named hello_world.s, containing the generated assembly instructions. On Mac OS 10.10.4, where cc is an alias for clang, the following output is generated:

```
.section__TEXT,__text,regular,pure_instructions
.macosx_version_min 10, 10
.globl_main
.align4, 0x90
_main:## @main
.cfi_startproc
## BB#0:
pushq %rbp
Ltmp0:
.cfi_def_cfa_offset 16
Ltmp1:
.cfi_offset %rbp, -16
movq%rsp, %rbp
Ltmp2:
.cfi_def_cfa_register %rbp
subq$16, %rsp
leaqL_.str(%rip), %rdi
movl$0, -4(%rbp)
callq _puts
xorl%ecx, %ecx
movl%eax, -8(%rbp)## 4-byte Spill
movl%ecx, %eax
addq$16, %rsp
popq%rbp
retq
.cfi_endproc
.section__TEXT,__cstring,cstring_literals
L_.str: ## @.str
.asciz"Hello, World!"

.subsections_via_symbols
```

Assembly

- During the assembly stage, an assembler is used to translate the assembly instructions to machine code, or object code. The output consists of actual instructions to be run by the target processor.

To save the result of the assembly stage, pass the -c option to cc:

```
cc -c hello_world.c
```

- Running the above command will create a file named hello_world.o, containing the object code of the program. The contents of this file is in a binary format and can be inspected using hexdump or od by running either one of the following commands:

```
hexdump hello_world.o
od -c hello_world.o
```

Linking

- The object code generated in the assembly stage is composed of machine instructions that the processor understands but some pieces of the program are out of order or missing.
- To produce an executable program, the existing pieces have to be rearranged and the missing ones filled in. This process is called linking.
- The linker will arrange the pieces of object code so that functions in some pieces can successfully call functions in other pieces.
- It will also add pieces containing the instructions for library functions used by the program.
- In the case of the —Hello, World!‖ program, the linker will add the object code for the puts function.

The result of this stage is the final executable program. When run without options, cc will name this file a.out. To name the file something else, pass the -o option to cc:

```
cc -o hello_world hello_world.c
```

UNIT II

ARRAYS AND STRINGS

Introduction to Arrays: Declaration, Initialization - One dimensional array - Example Program: Computing Mean, Median and Mode - Two dimensional arrays - Example Program: Matrix Operations (Addition, Scaling, Determinant and Transpose) - String operations: length, compare, concatenate, copy - Selection sort, linear and binary search.

2. INTRODUCTION TO ARRAYS

2.1. Overview

- An array is a collection of data items, all of the same type, accessed using a common name.
- A one-dimensional array is like a list;A two dimensional array is like a table;The C language places no limits on the number of dimensions in an array, though specific implementations may.
- Some texts refer to one-dimensional arrays as vectors, two-dimensional arrays as matrices, and use the general term arrays when the number of dimensions is unspecified or unimportant.
- All arrays consist of contiguous memory locations. The lowest address corresponds to the first element and the highest address to the last element.

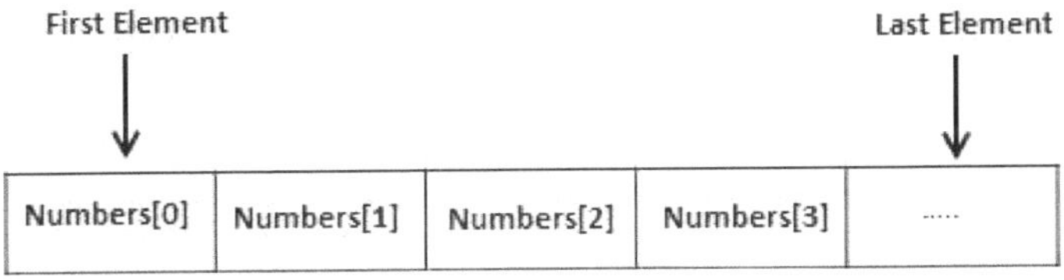

2.2. Array Declaration

- Array variables are declared identically to variables of their data type, except that the variable name is followed by one pair of square [ ] brackets for each dimension of the array.

- Uninitialized arrays must have the dimensions of their rows, columns, etc. listed within the square brackets.
- Dimensions used when declaring arrays in C must be positive integral constants or constant expressions.

Examples

```
int i, j, intArray[ 10 ], number;
float floatArray[ 1000 ];
int tableArray[ 3 ][ 5 ];/* 3 rows by 5 columns */
const int NROWS = 100;// ( Old code would use #define NROWS 100 )
const int NCOLS = 200;// ( Old code would use #define NCOLS 200 )
float matrix[ NROWS ][ NCOLS ];
```

2.3. Initializing Arrays

- Arrays may be initialized when they are declared, just as any other variables.
- Place the initialization data in curly {} braces following the equals sign.Note the use of commas in the examples below.
- An array may be partially initialized, by providing fewer data items than the size of the array.The remaining array elements will be automatically initialized to zero.
- If an array is to be completely initialized, the dimension of the array is not required.The compiler will automatically size the array to fit the initialized data. (Variation: Multi dimensional arrays - see below.)

Examples:

```
int i =5, intArray[ 6 ] = { 1, 2, 3, 4, 5, 6 }, k;
float sum= 0.0f, floatArray[ 100 ] = { 1.0f, 5.0f, 20.0f };
doublepiFractions[ ] = { 3.141592654, 1.570796327, 0.785398163 };
```

2.3.1. Using arrays

- Elements of an array are accessed by specifying the index (offset) of the desired element within square [ ] brackets after the array name.
- Array subscripts must be of integer type.(int, long int, char, etc.)
- **Very Important:** Array indices start at zero in C, and go to one less than the size of the array.For example, a five

element array will have indices zero through four.This is because the index in C is actually an offset from the beginning of the array.(The first element is at the beginning of the array, and hence has zero offset.)

- **Landmine:**The most common mistake when working with arrays in C is forgetting that indices start at zero and stop one less than the array size.
- Arrays are commonly used in conjunction with loops, in order to perform the same calculations on all (or some part) ofthe data items in the array.

2.3.2. Accessing array elements

An element is accessed by indexing the array name. This is done by placing the index of the element within square brackets after the name of the array.

Example

```
double salary = balance[9];
```

The above statement will take the 10th element from the array and assign the value to salary variable. The following example Shows how to use all the three above mentioned concepts viz. declaration, assignment, and accessing arrays

```
#include <stdio.h>

int main () {

 int n[ 10 ]; /* n is an array of 10 integers */
 int i,j;

 /* initialize elements of array n to 0 */
 for ( i = 0; i < 10; i++ ) {
n[ i ] = i + 100; /* set element at location i to i + 100 */
 }

 /* output each array element's value */
for (j = 0; j < 10; j++ ) {
printf("Element[%d] = %d\n", j, n[j] );
 }

 return 0;
}
```

Output:

```
Element[0] = 100
Element[1] = 101
Element[2] = 102
Element[3] = 103
Element[4] = 104
Element[5] = 105
Element[6] = 106
Element[7] = 107
Element[8] = 108
Element[9] = 109
```

2.4. One Dimensional Array

An array where data is arranged in a single dimension is called one dimensional array.

Syntax for declaration of One Dimensional Array

```
datatype array_name[size];
```

Here, array_name is an array of type datatype and size is the number of elements in that array.

Example:

```
int a[5]; // declares an integer array with 5 elements
float f[5]; // declares an float array with 5 elements
char c[10]; // declares an character array with 10 elements
```

2.4.1. Arrangement of one dimensional array

1	2	....	*n*

Figure: One Dimensional Array

Example of C Program to enter 5 numbers and print them in reverse order

```
#include<stdio.h>
int main()
{
int i,a[10];
printf("Enter 5 numbers\n");
for(i=0;i<5;i++)
```

```
{
scanf("%d",&a[i]);
}
printf("Reverse order\n");
for(i=4;i>=0;i--)
{
printf("%d\n",a[i]);
}
return 0;
}
```

- In this program, a one dimensional integer array with 5 elements is declared where these values are entered by the user.
- These values are then stored in respective index of array a using a for loop.
- Then, these values are printed in reverse order, i.e. the value with higher index is printed at first, by traversing through the array.

Output

```
Enter 5 numbers
11
34
-21
6
90
Reverse order
90
6
-21
34
11
```

2.5. Example Program

2.5.1. Example program of computing mean using one dimensional array

Following C program ask the user to enter "How many number he/she want to enter", then ask to enter all the numbers to calculate and display the arithmetic mean:

```
/* C Program - Calculate Arithmetic Mean */

#include<stdio.h>
#include<conio.h>
void main()
{
        clrscr();
        int n, i, arr[50], sum=0;
        printf("How many number you want to enter ?\n");
        scanf("%d",&n);
        printf("Enter %d Numbers :",n);
        for(i=0; i<n; i++)
        {
                scanf("%d",&arr[i]);
                sum=sum+arr[i];
        }
        int armean=sum/n;
        printf("Arithmetic Mean = %d",armean);
        getch();
}
```

Output:

```
How many number you want to enter?
10
Enter 10 numbers :
1
2
3
4
5
6
7
8
9
10
Arithmetic Mean: 5
```

2.5.2. Example program of Computing Average using one dimensional array.

```
#include <stdio.h>

int main()
{
int n, i;
float num[100], sum = 0.0, average;

printf("Enter the numbers of elements: ");
scanf("%d", &n);

while (n > 100 || n <= 0)
{
printf("Error! number should in range of (1 to 100).\n");
printf("Enter the number again: ");
scanf("%d", &n);
}

for(i = 0; i < n; ++i)
{
printf("%d. Enter number: ", i+1);
scanf("%f", &num[i]);
sum += num[i];
}

average = sum / n;
printf("Average = %.2f", average);

return 0;
}
```

Output:

```
Enter the numbers of elements: 6
1. Enter number: 45.3
2. Enter number: 67.5
3. Enter number: -45.6
4. Enter number: 20.34
5. Enter number: 33
6. Enter number: 45.6
Average = 27.69
```

- This program takes the number of elements in the array and stores in the variable n.
- Then, the for loop gets all the elements from the user and stores the sum of the entered numbers in sum.
- Finally, the average is calculated by dividing sum by the number of elements n.

2.5.3. Example program of C program to find the mean, median and mode of a set of numbers

```
#include<stdio.h>
main()
{
int i,j,a[20]={0},sum=0,n,t,b[20]={0},k=0,c=1,max=0,mode;
float x=0.0,y=0.0;
printf("\nEnter the limit\n");
scanf("%d",&n);
printf("Enter the set of numbers\n");
for(i=0;i<n;i++)
{
scanf("%d",&a[i]);
sum=sum+a[i];
}
x=(float)sum/(float)n;
printf("Mean\t= %f",x);

for(i=0;i<n;i++)
{
for(j=i+1;j<n;j++)
{
if(a[i]>a[j])
{ t=a[i];
a[i]=a[j];
a[j]=t;
}
}
}
```

```
if(n%2==0)
y=(float)(a[n/2]+a[(n-1)/2])/2;
else
y=a[(n-1)/2];
printf("\nMedian\t= %f",y);

for(i=0;i<n-1;i++)
{ mode=0;
for(j=i+1;j<n;j++)
{
if(a[i]==a[j])
{
mode++;
}
}
if((mode>max)&&(mode!=0))
{ k=0;
max=mode;
b[k]=a[i];
k++;
}
else if(mode==max)
{
b[k]=a[i];
k++;
}
}
for(i=0;i<n;i++)
{
if(a[i]==b[i])
c++;
}
if(c==n)
printf("\nThere is no mode");
else
```

```
{
printf("\nMode\t= ");
for(i=0;i<k;i++)
printf("%d ",b[i]);
}
}
```

Output:

```
Enter the limit
5
Enter the set of numbers
1 2 3 4 5
Mean=3.000000
Median=3.000000
There is no mode
Enter the limit
5
Enter the set of numbers
1 1 2 3 4
Mean=2.200000
Median=2.000000
Mode=1
Enter the set of numbers
1 1 2 2 3
Mean=1.800000
Median=2.000000
Mode=1 2 _
```

2.5.4. Example of C program to find mean, median, variance and standard deviation for given set of numbers.

```
#include <stdio.h>
#include <stdlib.h>
#include <math.h>
int main() {
float *x, mean = 0, median, sd, var;
int n, i, j, temp;
printf("Enter the no of entries:");
scanf("%d", &n);
```

```
x = (float *)malloc(sizeof (float) * n);

/* get n inputs from user */
printf("Enter your inputs:\n");
for (i = 0; i < n; i++)
scanf("%f", &x[i]);

/* calculate the mean */
for (i = 0; i < n; i++)
mean = mean + x[i];
mean = mean / n;

/* calculate the variance*/
for (i = 0; i < n; i++)
var = var + pow((x[i] - mean) , 2);

var = var / n;

/* square root of variance is SD */
sd= sqrt(var);

/* sort the given inputs to find median */
for (i = 0; i < n - 1; i++)
for (j = i; j < n; j++) {
if (x[i] > x[j]) {
temp = x[i];
x[i] = x[j];
x[j] = temp;
}
}

/* calculate the median */ if
((n + 1) % 2 == 0) { median
= x[((n + 1) / 2) - 1];
} else {
median = (x[((n + 1) / 2) - 1] + x[((n + 2) / 2) - 1]) / 2;
}
```

```
/* print the outputs */
printf("Standard deviation: %f\n", sd);
printf("Variance: %f\n", var);
printf("Mean : %f\n", mean);
printf("Median: %f\n", median);
return 0;
}
```

Output:

```
Enter the no of entries:5
Enter your inputs:
600
470
170
430
300
Standard deviation: 147.321609
Variance: 21703.656250
Mean : 394.000000
Median: 430.000000
```

2.6. Two Dimensional Arrays in C

- The two dimensional array in C language is represented in the form of rows and columns, also known as matrix. It is also known as array of arrays or list of arrays.
- The two dimensional, three dimensional or other dimensional arrays are also known as multidimensional arrays.

2.6.1. Declaration of two dimensional Array in C

We can declare an array in the c language in the following way.

```
data_type array_name[size1][size2];
```

A simple example to declare two dimensional array is given below.

```
int twodimen[4][3];
```

Here, 4 is the row number and 3 is the column number.

2.6.2. Initialization of 2D Array in C

A way to initialize the two dimensional array at the time of declaration is given below.

```
int arr[4][3]={{1,2,3},{2,3,4},{3,4,5},{4,5,6}};
```

Two dimensional array example in C

```
#include <stdio.h>
#include <conio.h>
void main(){
int i=0,j=0;
int arr[4][3]={{1,2,3},{2,3,4},{3,4,5},{4,5,6}};
clrscr();

//traversing 2D array
for(i=0;i<4;i++){
for(j=0;j<3;j++){
 printf("arr[%d] [%d] = %d \n",i,j,arr[i][j]);
 }//end of j
}//end of i

getch();
}
```

Output

```
arr[0][0] = 1
arr[0][1] = 2
arr[0][2] = 3
arr[1][0] = 2
arr[1][1] = 3
arr[1][2] = 4
arr[2][0] = 3
arr[2][1] = 4
arr[2][2] = 5
arr[3][0] = 4
arr[3][1] = 5
arr[3][2] = 6
```

2.6.3. Example program: Addition using 2D array in C programming

```
#include <stdio.h>
int main(){
int r, c, a[100][100], b[100][100], sum[100][100], i, j;

printf("Enter number of rows (between 1 and 100): ");
scanf("%d", &r);
printf("Enter number of columns (between 1 and 100): ");
scanf("%d", &c);

printf("\nEnter elements of 1st matrix:\n");

for(i=0; i<r; ++i)
for(j=0; j<c; ++j)
{
printf("Enter element a%d%d: ",i+1,j+1);
scanf("%d",&a[i][j]);
}

printf("Enter elements of 2nd matrix:\n");
for(i=0; i<r; ++i)
for(j=0; j<c; ++j)
{
printf("Enter element a%d%d: ",i+1, j+1);
scanf("%d", &b[i][j]);
}

// Adding Two matrices

for(i=0;i<r;++i)
for(j=0;j<c;++j)
{
sum[i][j]=a[i][j]+b[i][j];
}

// Displaying the result
printf("\nSum of two matrix is: \n\n");
```

```
for(i=0;i<r;++i)
for(j=0;j<c;++j)
{

printf("%d ",sum[i][j]);

if(j==c-1)
{
printf("\n\n");
}
}

return 0;
}
```

Output

```
Enter number of rows (between 1 and 100): 2
Enter number of columns (between 1 and 100): 3

Enter elements of 1st matrix:
Enter element a11: 2
Enter element a12: 3
Enter element a13: 4
Enter element a21: 5
Enter element a22: 2
Enter element a23: 3
Enter elements of 2nd matrix:
Enter element a11: -4
Enter element a12: 5
Enter element a13: 3
Enter element a21: 5
Enter element a22: 6
Enter element a23: 3

Sum of two matrix is:

-2 8 7

10 8 6
```

2.6.4. Example program: Scaling program using 2 dimensional array in C programming

```
#include<stdio.h>
#include<conio.h>
#include<graphics.h>
#include<process.h>
#include<math.h>

int x1, y1, x2, y2, x3, y3, mx, my;

void draw();
void scale();

void main() {
 int gd = DETECT, gm;

 initgraph(&gd, &gm, " ");
 printf("Enter the 1st point for the triangle:");
 scanf("%d%d", &x1, &y1);
 printf("Enter the 2nd point for the triangle:");
 scanf("%d%d", &x2, &y2);

 printf("Enter the 3rd point for the triangle:");
 scanf("%d%d", &x3, &y3);

 draw();

 scale();
}

void draw() {
 line(x1, y1, x2, y2);
 line(x2, y2, x3, y3);
 line(x3, y3, x1, y1);
}

void scale() {
```

```
int x, y, a1, a2, a3, b1, b2, b3;
int mx, my;

printf("Enter the scalling coordinates");
scanf("%d%d", &x, &y);

mx = (x1 + x2 + x3) / 3;
my = (y1 + y2 + y3) / 3;

cleardevice();

a1 = mx + (x1 - mx) * x;
b1 = my + (y1 - my) * y;
a2 = mx + (x2 - mx) * x;

b2 = my + (y2 - my) * y;
a3 = mx + (x3 - mx) * x;
b3 = my + (y3 - my) * y;

line(a1, b1, a2, b2);
line(a2, b2, a3, b3);
line(a3, b3, a1, b1);

draw();
getch();
}
```

2.6.5. Example of determinant program using 2D array in C programming

Write a C program to read elements in a matrix and find determinant of the given matrix. C program to find determinant of a 2x2 matrix and 3x3 matrix. Logic to find determinant of a matrix in C programming.

Example

Input

Input elements in 2x2 matrix:

1 2

3 4

Output

Determinant of the matrix = -2

What is Determinant of a Matrix?

- The Determinant of a matrix is a special number that can be calculated from the elements of a square matrix. The determinant of a matrix A is denoted by det (A), det A or |A|.
- Determinant of a Matrix is a special number that is defined only for square matrices (matrices which have same number of rows and columns). Determinant is used at many places in calculus and other matrix related algebra, it actually represents the matrix in term of a real number which can be used in solving system of linear equation and finding the inverse of a matrix.

$$A = \begin{bmatrix} a & b \\ c & d \end{bmatrix} \quad |A| = ad - bc$$

Determinant of 2x2 matrix

$$A = \begin{bmatrix} a & b & c \\ d & e & f \\ g & h & i \end{bmatrix} \quad |A| = a(ei - fh) - b(di - fg) + c(dh - eg)$$

Determinant of 3x3 matrix

```
/**
 * C program to find determinant of 2x2 matrix
 */

#include <stdio.h>
#define SIZE 2 // Matrix size

int main()
{
int A[SIZE][SIZE];
int row, col;
```

```
long det;
/* Input elements in matrix A from user */
printf("Enter elements in matrix of size 2x2: \n");
for(row=0; row<SIZE; row++)
{
for(col=0; col<SIZE; col++)
{
scanf("%d", &A[row][col]);
}
}

/*
 * det(A) = ad - bc
 * a = A[0][0], b = A[0][1], c = A[1][0], d = A[1][1]
 */
det = (A[0][0] * A[1][1]) - (A[0][1] * A[1][0]);

printf("Determinant of matrix A = %ld", det);

return 0;
}
```

Output:

```
Enter elements in matrix of size 2x2:
1 2
3 4
Determinant of matrix A = -2
```

2.6.6. C program to find transpose of a matrix using 2D Array

Write a C program to read elements in a matrix and find transpose of the given matrix. How to find transpose of a given matrix in C. Logic to find transpose of a matrix in C programming.

Example

Input

Input elements in matrix:

1 2 3

4 5 6

7 8 9

Output

Transpose:
1 4 7
2 5 8
3 6 9

Transpose of a matrix

Transpose of a matrix A is defined as converting all rows into columns and columns into rows. Transpose of matrix A is written as A^T.

$$\begin{bmatrix} 1 & 2 & 3 \\ 4 & 5 & 6 \\ 7 & 8 & 9 \end{bmatrix}^T = \begin{bmatrix} 1 & 4 & 7 \\ 2 & 5 & 8 \\ 3 & 6 & 9 \end{bmatrix}$$

Transpose of a matrix

Transpose of a matrix **A** is defined as
$\mathbf{A^T}_{ij}$; $\mathbf{A}_{ij}$ Where $1 \leq i \leq m$ and $1 \leq j \leq n$ 1

Logic to find transpose of a matrix

Below is the step by step descriptive logic to find transpose of a matrix.

- Input elements in matrix A from user.
- Declare another matrix of same size as of A, to store transpose of matrix say B.
- To iterate through each element of matrix run two loops. Run an outer loop from 0 to MAX_ROWS to iterate through rows. The loop structure should look like for(row=0; row<MAX_ROWS; row++).
- To iterate through each column of the matrix, run an inner loop from 0 to MAX_COLS. The loop structure must look like for(col=0; col<MAX_COLS; col++).
- Inside inner loop we will perform actual transpose of the matrix. As per definition column of transpose matrix is equal to row of original matrix and vice versa. Means assign B[col][row] = A[row][col].

```
/**
 * C program to find transpose of a matrix
 */

#include <stdio.h>
#define MAX_ROWS 3
#define MAX_COLS 3

int main()
{
int A[MAX_ROWS][MAX_COLS];// Original matrix
int B[MAX_COLS][MAX_ROWS];// Transpose matrix

int row, col;

/* Input elements in matrix A from user */
printf("Enter elements in matrix of size %dx%d: \n", MAX_ROWS,
MAX_COLS);
for(row=0; row<MAX_ROWS; row++)
{
for(col=0; col<MAX_COLS; col++)
{
scanf("%d", &A[row][col]);
}
}

/*
 * Find transpose of matrix A
 */
for(row=0; row<MAX_ROWS; row++)
{
for(col=0; col<MAX_COLS; col++)
{
/* Store each row of matrix A to each column of matrix B */
B[col][row] = A[row][col];
}
}

/* Print the original matrix A */
printf("\nOriginal matrix: \n");
for(row=0; row<MAX_ROWS; row++)
```

```
{
for(col=0; col<MAX_COLS; col++)
{
printf("%d ", A[row][col]);
}

printf("\n");
}

/* Print the transpose of matrix A */
printf("Transpose of matrix A: \n");
for(row=0; row<MAX_COLS; row++)
{
for(col=0; col<MAX_ROWS; col++)
{
printf("%d ", B[row][col]);
}

printf("\n");
}

return 0;
}
```

Output

```
Enter elements in matrix of size 3x3:
1 2 3
4 5 6
7 8 9

Original matrix:
1 2 3
4 5 6
7 8 9
Transpose of matrix A:
1 4 7
2 5 8
3 6 9
```

2.7. Strings And String Functions With Examples IN C

- String is an array of characters. In this guide, we learn how to declare strings, how to work with strings in C programming and how to use the pre-defined string handling functions.
- We will see how to compare two strings, concatenate strings, copy one string to another & perform various string manipulation operations. We can perform such operations using the pre-defined functions of —string.h‖ header file. In order to use these string functions you must include string.h file in your C program.

2.7.1. String Declaration

```
String Declaration
1) char str1[]={'A', 'B', 'C', 'D', '\0'};
2) char str1[]="ABCD";

        '\0' would automatically
        insterted at the end in this
        type of declaration
```

Method 1:

char address[]={'T', 'E', 'X', 'A', 'S', '\0'};

Method 2: The above string can also be defined as

char address[]="TEXAS";

In the above declaration NULL character (\0) will automatically be inserted at the end of the string.

What is NULL Char "\0"?

'\0' represents the end of the string. It is also referred as String terminator & Null Character.

2.7.2. String I/O in C programming

String I/O
1) printf and scanf
2) puts and gets
Syntax of above functions - Assume string as str1

```
printf("%s", str1);
puts(str1);  --%s not require here.
scanf("%s", &str1);
gets(str1); --%s not require
```

2.7.3. Read & write Strings in C using Printf() and Scanf() functions

```
#include <stdio.h>
#include <string.h>
int main()
{
/* String Declaration*/
char nickname[20];

printf("Enter your Nick name:");

/* I am reading the input string and storing it in nickname
 * Array name alone works as a base address of array so
 * we can use nickname instead of &nickname here
 */
scanf("%s", nickname);

/*Displaying String*/
printf("%s",nickname);

return 0;
}
```

Output

```
Enter your Nick name:Negan
Negan
```

Note: %s format specifier is used for strings input/output

2.7.4. Read & write strings in C using gets() and puts() functions

```
#include <stdio.h>
#include <string.h>
int main()
{
/* String Declaration*/
char nickname[20];

/* Console display using puts */
puts("Enter your Nick name:");

/*Input using gets*/
gets(nickname);

puts(nickname);

return 0;
}
```

2.7.5. C – String functions

strlen - Finds out the length of a string
strlwr - It converts a string to lowercase
strupr - It converts a string to uppercase
strcat - It appends one string at the end of another
strncat - It appends first n characters of a string at the end of another.
strcpy - Use it for Copying a string into another
strncpy - It copies first n characters of one string into another
strcmp - It compares two strings
strncmp - It compares first n characters of two strings
strcmpi - It compares two strings without regard to case ("i" denotes that this function ignores case)
stricmp - It compares two strings without regard to case (identical to strcmpi)
strnicmp - It compares first n characters of two strings, Its not case sensitive
strdup - Used for Duplicating a string
strchr - Finds out first occurrence of a given character in a string
strrchr - Finds out last occurrence of a given character in a string
strstr - Finds first occurrence of a given string in another string
strset - It sets all characters of string to a given character
strnset - It sets first n characters of a string to a given character
strrev - It Reverses a string

2.7.6. C string length

C String function - strlen

Syntax:

*size_t strlen(const char *str)*

size_t represents unsigned short

It returns the length of the string without including end character (terminating char ‗\0').

Example of strlen

```
#include <stdio.h>
#include <string.h>
int main()
{
 char str1[20] = "BeginnersBook";
 printf("Length of string str1: %d", strlen(str1));
 return 0;
}
```

Output

```
Length of string str1: 13
```

2.7.7. strlen vs sizeof

strlen returns you the length of the string stored in array, however sizeof returns the total allocated size assigned to the array. So if I consider the above example again then the following statements would return the below values.

- **strlen(str1)** returned value 13.
- **sizeof(str1)** would return value 20 as the array size is 20 (see the first statement in main function).

2.7.8. C String function - strnlen

Syntax:

*size_t strnlen(const char *str, size_t maxlen)*

- **size_t** represents unsigned short
- It returns length of the string if it is less than the value specified for maxlen (maximum length) otherwise it returns maxlen value.

Example of strnlen

```
#include <stdio.h>
#include <string.h>
int main()
{
 char str1[20] = "BeginnersBook";
 printf("Length of string str1 when maxlen is 30: %d", strnlen(str1, 30));
 printf("Length of string str1 when maxlen is 10: %d", strnlen(str1, 10));
 return 0;
}
```

Output

```
Length of string str1 when maxlen is 30: 13
Length of string str1 when maxlen is 10: 10
```

Have you noticed the output of second printf statement, even though the string length was 13 it returned only 10 because the maxlen was 10.

2.7.9. C String function - strcmp

Syntax

*int strcmp(const char *str1, const char *str2)*

It compares the two strings and returns an integer value. If both the strings are same (equal) then this function would return 0 otherwise it may return a negative or positive value based on the comparison.

- **If string1 < string2 OR string1** is a substring of string2 then it would result in a negative value. If string1 > string2 then it would return positive value.
- **If string1 == string2** then you would get 0(zero) when you use this function for compare strings.

Example of strcmp

```
#include <stdio.h>
#include <string.h>
int main()
{
 char s1[20] = "BeginnersBook";
```

```
char s2[20] = "BeginnersBook.COM";
if (strcmp(s1, s2) ==0)
{
printf("string 1 and string 2 are equal");
}else
{
printf("string 1 and 2 are different");
}
return 0;
}
```

Output

```
string 1 and 2 are different
```

2.7.10. C String function - strncmp

Syntax

*int strncmp(const char *str1, const char *str2, size_t n)*

- **size_t** is for unassigned short
- It compares both the string till n characters or in other words it compares first n characters of both the strings.

Example of strncmp:

```
#include <stdio.h>
#include <string.h>
int main()
{
char s1[20] = "BeginnersBook";
char s2[20] = "BeginnersBook.COM";
/* below it is comparing first 8 characters of s1 and s2*/
if (strncmp(s1, s2, 8) ==0)
{
printf("string 1 and string 2 are equal");
}else
{
printf("string 1 and 2 are different");
}
return 0;
}
```

Output

```
string1 and string 2 are equal
```

2.7.11. C String function - strcat

Syntax

*char *strcat(char *str1, char *str2)*

- It concatenates two strings and returns the concatenated string.

Example of strcat

```
#include <stdio.h>
#include <string.h>
int main()
{
 char s1[10] = "Hello";
 char s2[10] = "World";
 strcat(s1,s2);
 printf("Output string after concatenation: %s", s1);
 return 0;
}
```

Output:

```
Output string after concatenation: HelloWorld
```

2.7.12. C String function - strncat

Syntax:

```
char *strncat(char *str1, char *str2, int n)
```

- It concatenates n characters of str2 to string str1. A terminator char (‗\0‘) will always be appended at the end of the concatenated string.

Example of strncat:

```
#include <stdio.h>
#include <string.h>
int main()
{
 char s1[10] = "Hello";
 char s2[10] = "World";
 strncat(s1,s2, 3);
 printf("Concatenation using strncat: %s", s1);
 return 0;
}
```

Output:

```
Concatenation using strncat: HelloWor
```

2.7.13. C String function - strcpy

Syntax

*char *strcpy(char *str1, char *str2)*

- It copies the string str2 into string str1, including the end character (terminator char ‗\0‘).

Example of strcpy:

```
#include <stdio.h>
#include <string.h>
int main()
{
 char s1[30] = "string 1";
 char s2[30] = "string 2 : I"m gonna copied into s1";
 /* this function has copied s2 into s1*/
 strcpy(s1,s2);
 printf("String s1 is: %s", s1);
 return 0;
}
```

Output

```
String s1 is: string 2: I"m gonna copied into s1
```

2.7.14. C String function - strncpy

Syntax

*char *strncpy(char *str1, char *str2, size_t n)*

size_t is unassigned short and n is a number.

Case1: If length of str2 > n then it just copies first n characters of str2 into str1.

Case2: If length of str2 < n then it copies all the characters of str2 into str1 and appends several terminator chars(‗\0‘) to accumulate the length of str1 to make it n.

Example of strncpy

```
#include <stdio.h>
#include <string.h>
int main()
{
 char first[30] = "string 1";
 char second[30] = "string  2:I"m using strncpy now";
 /* this function has copied first 10 chars of s2 into s1*/
 strncpy(s1,s2, 12);
 printf("String s1 is: %s", s1);
 return 0;
}
```

Output:

```
String  s1 is: string 2:I"m
```

2.7.15. C String function - strchr

Syntax

*char *strchr(char *str, int ch)*

- It searches string str for character ch (you may be wondering that in above definition I have given data type of ch as int, don't worry I didn't make any mistake it should be int only.
- The thing is when we give any character while using strchr then it internally gets converted into integer for better searching.

Example of strchr:

```
#include <stdio.h>
#include <string.h>
int main()
{
 char mystr[30] = "I"m an example of function  strchr";
 printf ("%s", strchr(mystr, 'f'));
 return 0;
}
```

Output

```
f function strchr
```

2.7.16. C String function - Strrchr

Syntax

*char *strrchr(char *str, int ch)*

- It is similar to the function strchr, the only difference is that it searches the string in reverse order, now you would have understood why we have extra r in strrchr, yes you guessed it correct, it is for reverse only.

Now let's take the same above example:

```
#include <stdio.h>
#include <string.h>
int main()
{
 char mystr[30] = "I'm an example of function strchr";
 printf ("%s", strrchr(mystr, 'f'));
 return 0;
}
```

Output

```
function strchr
```

Why output is different than strchr? It is because it started searching from the end of the string and found the first _f' in function instead of _of'.

2.7.17. C String function - strstr

Syntax

*char *strstr(char *str, char *srch_term)*

- It is similar to strchr, except that it searches for string srch_term instead of a single char.

Example of strstr

```
#include <stdio.h>
#include <string.h>
int main()
{
 char inputstr[70] = "String Function in C at BeginnersBook.COM";
```

```
printf ("Output string is: %s", strstr(inputstr, 'Begi'));
return 0;
}
```

Output

```
Output string is: BeginnersBook.COM
```

You can also use this function in place of strchr as you are allowed to give single char also in place of search_term string.

2.8. Example of C Program to Perform String Operations

```
#include <stdio.h>
#include<conio.h>
#include<string.h>
#include<ctype.h>
#include<stdlib.h>

int length(char str[]);
void reverse(char str[]);
int palindrome(char str[]);
void copy(char str1[], char str2[]);
int compare(char str1[], char str2[]);
void concat(char str1[], char str2[]);
void search(char str1[], char str2[]);
void count(char str1[]);

void main() {
 char a[100], b[100];
 int result, option;
 do {
printf("\n1.Length of a string");
printf("\n2.Reverse the Given String");
printf("\n3.Check for Palindrome");
printf("\n4.Copy");
printf("\n5.String Comparison");
printf("\n6.String Concatenation");
printf("\n7.String Searching");
printf("\n8.Counting of Words,Characters & Special Characters");
printf("\n9.Quit");
```

```
printf("\n\nEnter Your Choice:");
scanf("%d", &option);

flushall();
switch (option) {
case 1:
 printf("\nEnter a String:");
 gets(a);
 result = length(a);
 printf("\nLength of %s=%d", a, result);
 printf("\nPress a Character");
 getch();
 break;

case 2:
 printf("\nEnter a String:");
 gets(a); reverse(a);
 printf("\nResult=%s", a);
 printf("\nPress a Character");
 getch();
 break;

case 3:
 printf("\n Enter a String:");
 gets(a);
 result = palindrome(a);
 if (result == 0)
printf("\nNot a palindrome");
 else
printf("\nA palindrome");
 printf("\nPress a Character");
 getch();
 break;

case 4:
 printf("\nEnter a String:");
```

```
gets(a); copy(b, a);
printf("\nResult=%s", b);
printf("\nPress a Character");
getch();
break;

case 5:
printf("\nEnter 1st string:");
gets(a);
printf("\nEnter 2nd string:");
gets(b);
result = compare(a, b); if
(result == 0)
printf("\nboth are same");
else if (result > 0)
printf("\n1st>2nd");
else printf("\n1st<2nd");
printf("\nPress a Character");
getch();
break;

case 6:
printf("\nEnter 1st string:");
gets(a);
printf("\nEnter 2nd string:");
gets(b); concat(a, b);
printf("\nresult=%s", a);
printf("\nPress a Character");
getch();
break;

case 7:
printf("\nEnter 1st string:");
gets(a);
```

```
 printf("\nEnter 2nd string:");
 gets(b);
 search(a, b);
 printf("\nPress a Character");
 getch();
 break;

case 8:
 printf("\nEnter a string:");
 gets(a);
 count(a);
 printf("\nPress a Character");
 getch();
 break;

default:
 printf("\nInvalid Choice:");
 break;
}

 } while (option != 9);
}

int length(char str[]) {
 int i = 0;
 while (str[i] != '\0')
i++;
 return (i);
}

void reverse(char str[]) {
 int i, j;
 char temp;
 i = j = 0;
 while (str[j] != '\0')
j++;
 j--;
```

```
 while (i < j) {
temp = str[i];
str[i] = str[j];
str[j] = temp;
i++;
j--;
 }
}

int palindrome(char str[]) {
 int i, j;
 i = j = 0;
 while (str[j] != '\0')
j++;
 while (i < j) {
if (str[i] != str[j - 1])
 return (0);
i++;
j--;
 }
 return (1);
}

void copy(char str2[], char str1[]) {
 int i = 0;
 while (str1[i] != '\0') {
str2[i] = str1[i];
i++;
 }
 str2[i] = '\0';
}

int compare(char str1[], char str2[]) {
 int i;
 i = 0;
 while (str1[i] != '\0') {
if (str1[i] > str2[i])
```

```
 return (1);
if (str1[i] < str2[i])
 return (-1);
i++;
 }
 return (0);
}

void concat(char str1[], char str2[]) {
 int i, j;
 i = 0;
 while (str1[i] != '\0')
i++;
 for (j = 0; str2[j] != '\0'; i++, j++)
str1[i] = str2[j];
 str1[i] = '\0';
}

void search(char str1[], char str2[]) {
 int i, j, lena, lenb;
 for (lena = 0; str1[lena] != '\0'; lena++);
 for (lenb = 0; str2[lenb] != '\0'; lenb++);

 for (i = 0; i <= lena - lenb + 1; i++) {
for (j = 0; str1[i + j] == str2[j] && str2[j] != '\0'; j++);
if (str2[j] == '\0')
 printf("\nString found at location : %d", i + 1);
 }
}

void count(char str[]) {
 int words = 0, characters = 0, spchar = 0, i;
 for (i = 0; str[i] != '\0'; i++) {
if (isalnum(str[i]) && (i == 0 || !isalnum(str[i - 1])))
 words++;
characters++;
if (!isalnum(str[i]) && !isspace(str[i]))
```

```
spchar++;
}
printf("\nNo of characters = %d", characters);
printf("\nNo of special characters = %d", spchar);
printf("\nNo of words = %d", words);
}
```

Output

```
1.Length of a string
2.Reverse the Given String
3.Check for Palindrome
4.Copy
5.String Comparison
6.String Concatenation
7.String Searching
8.Counting of Words,Characters & Special Characters
9.Quit

Enter Your Choice:1
Enter a String:pritesh
Length of pritesh=7
Press a Character

1.Length of a string
2.Reverse the Given String
3.Check for Palindrome
4.Copy
5.String Comparison
6.String Concatenation
7.String Searching
8.Counting of Words,Characters & Special Characters
9.Quit

Enter Your Choice:2
Enter a String:pritesh
Result = hsetirp
Press a Character
```

```
1.Length of a string
2.Reverse the Given String
3.Check for Palindrome
4.Copy
5.String Comparison
6.String Concatenation
7.String Searching
8.Counting of Words,Characters & Special Characters
9.Quit

Enter Your Choice:3
Enter a String:madam
A palindrome
Press a Character

1.Length of a string
2.Reverse the Given String
3.Check for Palindrome
4.Copy
5.String Comparison
6.String Concatenation
7.String Searching
8.Counting of Words,Characters & Special Characters
9.Quit

Enter Your Choice:4
Enter a String:pritesh
Result=pritesh
Press a Character

1.Length of a string
2.Reverse the Given String
3.Check for Palindrome
4.Copy
5.String Comparison
6.String Concatenation
```

```
7.String Searching
8.Counting of Words,Characters & Special Characters
9.Quit

Enter Your Choice:5
Enter 1st string:pri
Enter 2nd string:pri
both are same
Press a Character

1.Length of a string
2.Reverse the Given String
3.Check for Palindrome
4.Copy
5.String Comparison
6.String Concatenation
7.String Searching
8.Counting of Words,Characters & Special Characters
9.Quit

Enter Your Choice:6
Enter 1st string:prite
Enter 2nd string:sh
result=pritesh
Press a Character

1.Length of a string
2.Reverse the Given String
3.Check for Palindrome
4.Copy
5.String Comparison
6.String Concatenation
7.String Searching
8.Counting of Words,Characters & Special Characters
9.Quit
```

```
Enter Your Choice:7
Enter 1st string:pritesh
Enter 2nd string:pri
String found at location : 1
Press a Character

1.Length of a string
2.Reverse the Given String
3.Check for Palindrome
4.Copy
5.String Comparison
6.String Concatenation
7.String Searching
8.Counting of Words,Characters & Special Characters
9.Quit

Enter Your Choice:8
Enter a string:pri@20110
No of characters = 9
No of special characters = 1
No of words = 2
Press a Character

1.Length of a string
2.Reverse the Given String
3.Check for Palindrome
4.Copy
5.String Comparison
6.String Concatenation
7.String Searching
8.Counting of Words,Characters & Special Characters
9.Quit

Enter Your Choice:9
```

2.9. String Operations: Selection Sort, Linear and Binary Searc

Selection Sort

The selection sort algorithm sorts an array by repeatedly finding the minimum element (considering ascending order) from unsorted part and putting it at the beginning. The algorithm maintains two subarrays in a given array.

- The subarray which is already sorted.
- Remaining subarray which is unsorted.

In every iteration of selection sort, the minimum element (considering ascending order) from the unsorted subarray is picked and moved to the sorted subarray.

Following example explains the above steps:

```
arr[] = 64 25 12 22 11

// Find the minimum element in arr[0...4]
// and place it at beginning
11 25 12 22 64

// Find the minimum element in arr[1...4]
// and place it at beginning of arr[1...4]
11 12 25 22 64

// Find the minimum element in arr[2...4]
// and place it at beginning of arr[2...4]
11 12 22 25 64

// Find the minimum element in arr[3...4]
// and place it at beginning of arr[3...4]
11 12 22 25 64
```

2.9.1. Example of c program for implementation of selection sort

```
// C program for implementation of selection sort
#include <stdio.h>

void swap(int *xp, int *yp)
{
```

```
int temp = *xp;
*xp = *yp;
*yp = temp;
}

void selectionSort(int arr[], int n)
{
int i, j, min_idx;

// One by one move boundary of unsorted subarray
for (i = 0; i < n-1; i++)
{
// Find the minimum element in unsorted array
min_idx = i;
for (j = i+1; j < n; j++)
if (arr[j] < arr[min_idx])
min_idx = j;

// Swap the found minimum element with the first element
swap(&arr[min_idx], &arr[i]);
}
}

/* Function to print an array */
void printArray(int arr[], int size)
{
int i;
for (i=0; i < size; i++)
printf("%d ", arr[i]);
printf("\n");
}

// Driver program to test above functions
int main()
{
int arr[] = {64, 25, 12, 22, 11};
int n = sizeof(arr)/sizeof(arr[0]);
```

```
selectionSort(arr, n);
printf("Sorted array: \n");
printArray(arr, n);
return 0;
}
```

Output

```
Sorted array:
11      2 22 25 64
```

2.9.2. Example of c program for implementation of selection sort selection sort in C

C program for selection sort to sort numbers. This code implements selection sort algorithm to arrange numbers of an array in ascending order. With a little modification it will arrange numbers in descending order.

Selection sort algorithm implementation in C

```
#include <stdio.h>

int main()
{
 int array[100], n, c, d, position, swap;

 printf("Enter number of elements\n");
 scanf("%d", &n);

 printf("Enter %d integers\n", n);

 for ( c = 0 ; c < n ; c++ )
scanf("%d", &array[c]);

 for ( c = 0 ; c < ( n - 1 ) ; c++ )
 {
position = c;

for ( d = c + 1 ; d < n ; d++ )
```

```
{
 if ( array[position] > array[d] )
position = d;
}
if ( position != c )
{
 swap = array[c];
 array[c] = array[position];
 array[position] = swap;
}
 }

 printf("Sorted list in ascending order:\n");

 for ( c = 0 ; c < n ; c++ )
printf("%d\n", array[c]);

 return 0;
}
```

Output

```
Enter number of elements
10
Enter 10 integers
12
8
-6
2
4
5
3
7
4
2
Sorted list in ascending order:
-6
2
```

```
2
3
4
5
7
8
12
```

2.10. String Operations: Linear and Binary Search in C

2.10.1. Introduction to Searching

- Searching means to find whether a particular value is present in an array or not. If the value is present in the array, then searching is said to be successful and the searching process gives the location of that value in the array.

- However, if the value is not present in the array, the searching process displays an appropriate message and in this case searching is said to be unsuccessful.

  There are two popular methods for searching the array elements:
- Linear Search
- Binary Search

The algorithm that should be used depends entirely on how the values are organized in the array.

For example, if the elements of the array are arranged in ascending order, then binary search should be used, as it is more efficient for sorted lists in terms of complexity.

2.10.2. Linear Search

- Linear search, also called as sequential search, is a very simple method used for searching an array for a particular value. It works by comparing the value to be searched with every element of the array one by one in a sequence until a match is found.

- Linear search is mostly used to search an unordered list of elements (array in which data elements are not sorted). For example, if an array A[10] is declared and initialized as,

```
int A[10] = {10, 8, 2, 7, 3, 4, 9, 1, 6, 5};
Val = 7thenPos = 3
```

Linear Search Algorithm

```
LINEAR_SEARCH(A, N, VAL)

Step 1: [INITIALIZE] SET POS = -1
Step 2: [INITIALIZE] SET I = 1
Step 3: Repeat Step 4 while I<=N
Step 4: IF A[I] = VAL
 SET POS = I
 PRINT POS
 Go to Step 6
[END OF IF]
SET I = I + 1
[END OF LOOP]
Step 5: IF POS = –1
PRINT VALUE IS NOT PRESENT IN THE ARRAY
[END OF IF]
Step 6: EXIT
```

2.10.3. Complexity of linear search algorithm

- Linear search executes in O(n) time where n is the number of elements in the array.
- Obviously, the best case of linear search is when VAL is equal to the first element of the array. In this case, only one comparison will be made.
- Likewise, the worst case will happen when either VAL is not present in the array or it is equal to the last element of the array. In both the cases, n comparisons will have to be made.

2.10.4. Binary search

- Binary search is a searching algorithm that works efficiently with a sorted list.
- The mechanism of binary search can be better understood by an analogy of a telephone directory. When we are searching for a particular name in a directory, we first open the directory from the middle and then decide whether to look for the name in the first part of the directory or in the second part of the directory. Again, we open some page in the middle and the whole process is repeated until we finally find the right name.

The same mechanism is applied in the binary search.

- Now, let us consider how this mechanism is applied to search for a value in a sorted array.
- Consider an array A[11] that is declared and initialized as,

```
int A[11] = {0, 1, 2, 3, 4, 5, 6, 7, 8, 9, 10};
and
value to be searched is VAL = 9
```

The algorithm will proceed in the following manner.

```
BEG = 0, END = 10, MID = (0 + 10)/2 = 5
Now, VAL = 9 and A[MID] = A[5] = 5

A[5] is less than VAL, therefore, we now search for the value in
the second half of the array. So, we change the values of BEG
and MID.

Now, BEG = MID + 1 = 6, END = 10, MID = (6 + 10)/2 =16/2 = 8
VAL = 9 and A[MID] = A[8] = 8

A[8] is less than VAL, therefore, we now search for the value in
the second half of the segment.
So, again we change the values of BEG and MID.

Now, BEG = MID + 1 = 9, END = 10, MID = (9 + 10)/2 = 9
Now, VAL = 9 and A[MID] = 9.
```

The algorithm will terminate when A[MID] = VAL. When the algorithm ends, we will set POS = MID. POS is the position at which the value is present in the array.

However, if VAL is not equal to A[MID], then the values of BEG, END, and MID will be changed depending on whether VAL is smaller or greater than A[MID].

(a) If VAL < A[MID], then VAL will be present in the left segment of the array. So, the value of END will be changed as END = MID – 1. (b) If VAL > A[MID], then VAL will be present in the right segment of the array. So, the value of BEG will be changed as BEG = MID + 1.

Binary Search Algorithm

BINARY_SEARCH(A, lower_bound, upper_bound, VAL)

Step 1: [INITIALIZE] SET BEG = lower_bound
END = upper_bound, POS = - 1
Step 2: Repeat Steps 3 and 4 while BEG <= END Step 3: SET MID = (BEG + END)/2 Step 4: IF A[MID] = VAL SET POS = MID PRINT POS Go to Step 6 ELSE IF A[MID] > VAL
SET END = MID - 1
ELSE
SET BEG = MID + 1
[END OF IF]
[END OF LOOP]
Step 5: IF POS = -1
PRINT —VALUE IS NOT PRESENT IN THE ARRAY‖
[END OF IF]
Step 6: EXIT

2.10.5. Linear search in C programming

- The following code implements linear search (Searching algorithm) which is used to find whether a given number is present in an array and if it is present then at what location it occurs.

- It is also known as sequential search.

It is very simple and works as follows:

We keep on comparing each element with the element to search until the desired element is found or list ends. Linear search in C language for multiple occurrences and using function.

Example of Linear search c program

```
#include <stdio.h>

int main()
{
 int array[100], search, c, n;

 printf("Enter the number of elements in array\n");
 scanf("%d",&n);

 printf("Enter %d integer(s)\n", n);

 for (c = 0; c < n; c++)
scanf("%d", &array[c]);

 printf("Enter the number to search\n");
 scanf("%d", &search);

 for (c = 0; c < n; c++)
 {
if (array[c] == search) /* if required element found */
{
 printf("%d is present at location %d.\n", search, c+1);
 break;
}
 }
 if (c == n)
printf("%d is not present in array.\n", search);
```

```
return 0;
}
```

Output

```
Enter the number of elements in array
5
Enter the 5 numbers
5
6
4
2
9
Enter the number to search
4
4 is the present location 3
```

2.10.6. C program for binary search

This code implements binary search in c language. It can only be used for sorted arrays, but it's fast as compared to linear search. If you wish to use binary search on an array which is not sorted then you must sort it using some sorting technique say merge sort and then use binary search algorithm to find the desired element in the list. If the element to be searched is found then its position is printed. The code below assumes that the input numbers are in ascending order.

Example of C programming code for binary search

```
#include <stdio.h>

int main()
{
 int c, first, last, middle, n, search, array[100];

 printf("Enter number of elements\n");
 scanf("%d",&n);
```

```
printf("Enter %d integers\n", n);

for (c = 0; c < n; c++)
scanf("%d",&array[c]);

printf("Enter value to find\n");
scanf("%d", &search);

first = 0;
last = n - 1;
middle = (first+last)/2;

while (first <= last) {
if (array[middle] < search)
first = middle + 1;
else if (array[middle] == search) {
printf("%d found at location %d.\n", search, middle+1);
break;
}
else
last = middle - 1;

middle = (first + last)/2;
}
if (first > last)
printf("Not found! %d is not present in the list.\n", search);

return 0;
}
```

Output

```
Enter number of elements
7
Enter 7 integers
-4
5
8
```

```
9
11
43
485
Enter value to find
11
11        ound at location 5
```

UNIT III
FUNCTIONS AND POINTERS

Introduction to functions: Function prototype, function definition, function call, Built-in functions (string functions, math functions) - Recursion - Example Program: Computation of Sine series, Scientific calculator using built-in functions, Binary Search using recursive functions - Pointers - Pointer operators - Pointer arithmetic - Arrays and pointers - Array of pointers - Example Program: Sorting of names - Parameter passing: Pass by value, Pass by reference - Example Program: Swapping of two numbers and changing the value of a variable using pass by reference

3. INTRODUCTION TO FUNCTION

- A function is a group of statements that together perform a task. Every C program has at least one function, which is main(), and all the most trivial programs can define additional functions.
- You can divide up your code into separate functions. How you divide up your code among different functions is up to you, but logically the division is such that each function performs a specific task.
- A function **declaration** tells the compiler about a function's name, return type, and parameters. A function definition provides the actual body of the function.
- The C standard library provides numerous built-in functions that your program can call. For example, **strcat()** to concatenate two strings, **memcpy()** to copy one memory location to another location, and many more functions.
- A function can also be referred as a method or a sub-routine or a procedure, etc.

Why we need functions in C?

Functions are used because of following reasons

- To improve the readability of code.
- Improves the reusability of the code, same function can be used in any program rather than writing the same code from scratch.

- Debugging of the code would be easier if you use functions, as errors are easy to be traced.
- Reduces the size of the code, duplicate set of statements are replaced by function calls.

3.1. C Function Prototype Declaration

Function prototype declaration in C Programming

- Function prototype declaration is necessary in order to provide information to the compiler about function, about return type, parameter list and function name etc.

Important Points

- Our program starts from main function. Each and every function is called directly or indirectly through main function
- Like variable we also need to declare function before using it in program.
- In C, declaration of function is called as prototype declaration
- Function declaration is also called as function prototype

Points to remember

Below are some of the important notable things related to prototype declaration –

- It tells name of function,return type of function and argument list related information to the compiler
- Prototype declaration always ends with semicolon.
- Parameter list is optional.
- Default return type is integer.

Defining a Function

The general form of a function definition in C programming language is as follows –

```
return_type function_name( parameter list )
{
 body of the function
}
```

A function definition in C programming consists of a function header and a function body. Here are all the parts of a function.

Return Type: A function may return a value. The **return_type** is the data type of the value the function returns. Some functions perform the desired operations without returning a value. In this case, the return_type is the keyword **void.**

Function Name: This is the actual name of the function. The function name and the parameter list together constitute the function signature.

Parameters: A parameter is like a placeholder. When a function is invoked, you pass a value to the parameter. This value is referred to as actual parameter or argument. The parameter list refers to the type, order, and number of the parameters of a function. Parameters are optional; that is, a function may contain no parameters.

Function Body: The function body contains a collection of statements that define what the function does.

Example

Given below is the source code for a function called max(). This function takes two parameters num1 and num2 and returns the maximum value between the two –

```
/* function returning the max between two numbers */
int max(int num1, int num2) {

 /* local variable declaration */
 int result;

 if (num1 > num2)
     result = num1;
 else
 result = num2;

 return result;
}
```

Need of prototype declaration

- Program Execution always starts from main , but during lexical analysis (1st Phase of Compiler) token generation starts from left to right and from top to bottom.
- During code generation phase of compiler it may face issue of backward reference.

If we write prototype declaration then

- Prototype declaration tells compiler that we are going to define this function somewhere in the program.
- Compiler will have prior information about function.
- As compiler have prior information ,during function calling compiler looks forward in the program for the function definition.

If we don't write prototype declaration then

- Compilers don't have any reference of Function.
- Compilers don't have prior information of that function.
- Compiler gets confused and interprets it as unknown reference and throws error.

3.2. Function Declarations

A function declaration tells the compiler about a function name and how to call the function. The actual body of the function can be defined separately.

A function declaration has the following parts
return_type function_name(parameter list);

For the above defined function max(), the function declaration is as follows –
int max(int num1, int num2);

Parameter names are not important in function declaration only their type is required, so the following is also a valid declaration –
int max(int, int);

Function declaration is required when you define a function in one source file and you call that function in another file. In such case, you should declare the function at the top of the file calling the function.

3.3. Calling a Function

- While creating a C function, you give a definition of what the function has to do. To use a function, you will have to call that function to perform the defined task.
- When a program calls a function, the program control is transferred to the called function. A called function performs a defined task and when its return statement is executed or when its function-ending closing brace is reached, it returns the program control back to the main program.
- To call a function, you simply need to pass the required parameters along with the function name, and if the function returns a value, then you can store the returned value.

Example

```
#include <stdio.h>

/* function declaration */
int max(int num1, int num2);

int main () {

 /* local variable definition */
 int a = 100;
 int b = 200;
 int ret;

 /* calling a function to get max value */
 ret = max(a, b);

 printf( "Max value is : %d\n", ret );

 return 0;
}
```

```
/* function returning the max between two numbers */
int max(int num1, int num2) {

 /* local variable declaration */
 int result;

 if (num1 > num2)
result = num1;
 else
result = num2;

 return result;
}
```

We have kept max() along with main() and compiled the source code. While running the final executable, it would produce the following result.

```
Max value is : 200
```

3.4. Built in Function: C Library - <string.h>

The string.h header defines one variable type, one macro, and various functions for manipulating arrays of characters.

Library Variables

Following is the variable type defined in the header **string.h**

Variable & Description

- **size_t** - This is the unsigned integral type and is the result of the sizeof keyword.

Library Macros

Following is the macro defined in the header string.h

- **NULL** - This macro is the value of a null pointer constant.

Library Functions

Following are the functions defined in the header string.h

i. **C library function - memchr()**
   Description

- The C library function **void *memchr(const void *str, int c, size_t n)**
- Searches for the first occurrence of the character c (an unsigned char) in the first n bytes of the string pointed to, by the argument str.

Declaration

Following is the declaration for memchr() function.

void *memchr(const void *str, int c, size_t n)

Parameters

- **str** – This is the pointer to the block of memory where the search is performed.
- **c** – This is the value to be passed as an int, but the function performs a byte per byte search using the unsigned char conversion of this value.
- **n** – This is the number of bytes to be analyzed.

Return Value

This function returns a pointer to the matching byte or NULL if the character does not occur in the given memory area.

Example

The following example shows the usage of memchr() function.

```
#include <stdio.h>
#include <string.h>

int main () {
 const char str[] = "http://www.tutorialspoint.com";
 const char ch = '.';
 char *ret;

 ret = memchr(str, ch, strlen(str));

 printf("String after |%c| is - |%s|\n", ch, ret);
 return(0);
}
```

Let us compile and run the above program that will produce the following result.

```
String after |.| is - |.tutorialspoint.com|
```

ii. **C library function - memcmp()**

Description

- The C library function **int memcmp(const void *str1, const void *str2, size_t n))**
- Compares the first n bytes of memory area str1 and memory area str2.

Declaration

Following is the declaration for memcmp() function.

int memcmp(const void *str1, const void *str2, size_t n)

Parameters

- **str1** – This is the pointer to a block of memory.
- **str2** – This is the pointer to a block of memory.
- **n** – This is the number of bytes to be compared.

Return Value

- if Return value < 0 then it indicates str1 is less than str2.
- if Return value > 0 then it indicates str2 is less than str1.
- if Return value = 0 then it indicates str1 is equal to str2.

Example

The following example shows the usage of memcmp() function.

```
#include <stdio.h>
#include <string.h>

int main () {
 char str1[15];
 char str2[15];
 int ret;

 memcpy(str1, "abcdef", 6);
 memcpy(str2, "ABCDEF", 6);

 ret = memcmp(str1, str2, 5);

 if(ret > 0) {
```

```
printf("str2 is less than str1");
 } else if(ret < 0) {
printf("str1 is less than str2");
 } else {
printf("str1 is equal to str2");
 }

 return(0);
}
```

Let us compile and run the above program that will produce the following result.

```
str2 is less than str1
```

iii. C library function - memcpy()

Description

- The C library function void ***memcpy(void *str1, const void *str2, size_t n)**
- Copies n characters from memory area str2 to memory area str1.

Declaration

Following is the declaration for memcpy() function.

void *memcpy(void *str1, const void *str2, size_t n)

Parameters

- **str1** – This is pointer to the destination array where the content is to be copied, type-casted to a pointer of type void*.
- **str2** – This is pointer to the source of data to be copied, type-casted to a pointer of type void*.
- **n** – This is the number of bytes to be copied.

Return Value

This function returns a pointer to destination, which is str1.

Example

The following example shows the usage of memcpy() function.

```
#include <stdio.h>
#include <string.h>
```

```
int main () {
 const char src[50] = "http://www.studentspoint.com";
 char dest[50];

 printf("Before memcpy dest = %s\n", dest);
 memcpy(dest, src, strlen(src)+1);
 printf("After memcpy dest = %s\n", dest);

 return(0);
}
```

Let us compile and run the above program that will produce the following result –

```
Before memcpy dest =
After memcpy dest = http://www. studentspoint.com
```

iv. C library function - memmove()

Description

- The C library function **void *memmove(void *str1, const void *str2, size_t n)**
- Copies n characters from str2 to str1, but for overlapping memory blocks, memmove() is a safer approach than memcpy().

Declaration

Following is the declaration for memmove() function.

void *memmove(void *str1, const void *str2, size_t n)

Parameters

- **str1** – This is a pointer to the destination array where the content is to be copied, type-casted to a pointer of type void*.
- **str2** – This is a pointer to the source of data to be copied, type-casted to a pointer of type void*.
- **n** – This is the number of bytes to be copied.

Return Value

This function returns a pointer to the destination, which is str1.

Example

The following example shows the usage of memmove() function.

```
#include <stdio.h>
#include <string.h>

int main () {
 char dest[] = "oldstring";
 const char src[]= "newstring";

 printf("Before memmove dest = %s, src = %s\n", dest, src);
 memmove(dest, src, 9);
 printf("After memmove dest = %s, src = %s\n", dest, src);

 return(0);
}
```

Let us compile and run the above program that will produce the following result –

```
Before memmove dest = oldstring, src = newstring
After memmove dest = newstring, src = newstring
```

v. C library function - memset()

Description

- The C library function **void *memset(void *str, int c, size_t n)**
- Copies the character c (an unsigned char) to the first n characters of the string pointed to, by the argument str.

Declaration

Following is the declaration for memset() function.

void *memset(void *str, int c, size_t n)

Parameters

- **str** – This is a pointer to the block of memory to fill.
- **c** – This is the value to be set. The value is passed as an int, but the function fills the block of memory using the unsigned char conversion of this value.
- **n** – This is the number of bytes to be set to the value.

Return Value

This function returns a pointer to the memory area str.

Example

The following example shows the usage of memset() function.

```
#include <stdio.h>
#include <string.h>

int main () {
 char str[50];

 strcpy(str,"This is string.h library function");
 puts(str);

 memset(str,'$',7);
 puts(str);

 return(0);
}
```

Let us compile and run the above program that will produce the following result –

```
This is string.h library function
$$$$$$$ string.h library function
```

vi. **C library function - strcat()**

Description

- The C library function **char *strcat(char *dest, const char *src)**
- Appends the string pointed to by src to the end of the string pointed to by dest.

Declaration

Following is the declaration for strcat() function.

char *strcat(char *dest, const char *src)

Parameters

- **dest** – This is pointer to the destination array, which should contain a C string, and should be large enough to contain the concatenated resulting string.

- **src** – This is the string to be appended. This should not overlap the destination.

Return Value

This function returns a pointer to the resulting string dest.

Example

The following example shows the usage of strcat() function.

```
#include <stdio.h>
#include <string.h>

int main () {
 char src[50], dest[50];

 strcpy(src,"This is source");
 strcpy(dest, "This is destination");

 strcat(dest, src);

 printf("Final destination string : |%s|", dest);

 return(0);
}
```

Let us compile and run the above program that will produce the following result –

```
Final destination string : |This is destinationThis is source|
```

vii. C library function - strncat()

Description

- The C library function **char *strncat(char *dest, const char *src, size_t n)**
- Appends the string pointed to by src to the end of the string pointed to by dest up to n characters long.

Declaration

Following is the declaration for strncat() function.

char *strncat(char *dest, const char *src, size_t n)

Parameters

- **dest** – This is pointer to the destination array, which should contain a C string, and should be large enough to contain the concatenated resulting string which includes the additional null-character.
- **src** – This is the string to be appended.
- **n** – This is the maximum number of characters to be appended.

Return Value

This function returns a pointer to the resulting string dest.

Example

The following example shows the usage of strncat() function.

```
#include <stdio.h>
#include <string.h>

int main () {
 char src[50], dest[50];

 strcpy(src,"This is source");
 strcpy(dest, "This is destination");

 strncat(dest, src, 15);

 printf("Final destination string : |%s|", dest);

 return(0);
}
```

Let us compile and run the above program that will produce the following result –

```
Final destination string : |This is destinationThis is source|
```

viii. C library function - strchr()

Description

- The C library function char *strchr(const char *str, int c)
- Searches for the first occurrence of the character c (an unsigned char) in the string pointed to by the argument str.

Declaration

Following is the declaration for strchr() function.

char *strchr(const char *str, int c)

Parameters

- **str** – This is the C string to be scanned.
- **c** – This is the character to be searched in str.

Return Value

This returns a pointer to the first occurrence of the character c in the string str, or NULL if the character is not found.

Example

The following example shows the usage of strchr() function.

```
#include <stdio.h>
#include <string.h>

int main () {
 const char str[] = "http://www. studentspoint.com";
 const char ch = '.';
 char *ret;

 ret = strchr(str, ch);

 printf("String after |%c| is - |%s|\n", ch, ret);

 return(0);
}
```

Let us compile and run the above program that will produce the following result –

```
String after |.| is - |.studentspoint.com|
```

ix. C library function - strcmp()

Description

- The C library function **int strcmp(const char *str1, const char *str2)**
- Compares the string pointed to, by str1 to the string pointed to by str2.

Declaration

Following is the declaration for strcmp() function.

int strcmp(const char *str1, const char *str2)

Parameters

- **str1** – This is the first string to be compared.
- **str2** – This is the second string to be compared.

Return Value

This function return values that are as follows:

- if Return value < 0 then it indicates str1 is less than str2.
- if Return value > 0 then it indicates str2 is less than str1.
- if Return value = 0 then it indicates str1 is equal to str2.

Example

The following example shows the usage of strncmp() function.

```
#include <stdio.h>
#include <string.h>

int main () {
 char str1[15];
 char str2[15];
 int ret;


 strcpy(str1, "abcdef");
 strcpy(str2, "ABCDEF");

 ret = strcmp(str1, str2);

 if(ret < 0) {
printf("str1 is less than str2");
 } else if(ret > 0) {
printf("str2 is less than str1");
 } else {
printf("str1 is equal to str2");
 }

 return(0);
}
```

Let us compile and run the above program that will produce the following result.

```
str2 is less than str1
```

x. C library function - strncmp()

Description

- The C library function **int strncmp(const char *str1, const char *str2, size_t n)**
- Compares at most the first n bytes of str1 and str2.

Declaration

Following is the declaration for strncmp() function.

int strncmp(const char *str1, const char *str2, size_t n)

Parameters

- **str1** – This is the first string to be compared.
- **str2** – This is the second string to be compared.
- **n** – The maximum number of characters to be compared.

Return Value

These functions return values that are as follows.

- if Return value < 0 then it indicates str1 is less than str2.
- if Return value > 0 then it indicates str2 is less than str1.
- if Return value = 0 then it indicates str1 is equal to str2.

Example

The following example shows the usage of strncmp() function.

```
#include <stdio.h>
#include <string.h>

int main () {
 char str1[15];
 char str2[15];
 int ret;

 strcpy(str1, "abcdef");
 strcpy(str2, "ABCDEF");
```

```
ret = strncmp(str1, str2, 4);

if(ret < 0) {
printf("str1 is less than str2");
} else if(ret > 0) {
printf("str2 is less than str1");
} else {
printf("str1 is equal to str2");
}

return(0);
}
```

Let us compile and run the above program that will produce the following result –

```
str2 is less than str1
```

xi. C library function - strcat()

Description

- The C library function **char *strcat(char *dest, const char *src)**
- Aappends the string pointed to by src to the end of the string pointed to by dest.

Declaration

Following is the declaration for strcat() function.

char *strcat(char *dest, const char *src)

Parameters

- **dest** – This is pointer to the destination array, which should contain a C string, and should be large enough to contain the concatenated resulting string.
- **src** – This is the string to be appended. This should not overlap the destination.

Return Value

This function returns a pointer to the resulting string dest.

Example

The following example shows the usage of strcat() function.

```
#include <stdio.h>
#include <string.h>

int main () {
 char src[50], dest[50];

 strcpy(src,"This is source");
 strcpy(dest, "This is destination");

 strcat(dest, src);

 printf("Final destination string : |%s|", dest);

 return(0);
}
```

Let us compile and run the above program that will produce the following result −

```
Final destination string : |This is destinationThis is source|
```

xii. C library function - strcoll()

Description

- The C library function **int strcoll(const char *str1, const char *str2)**
- Compares string str1 to str2. The result is dependent on the LC_COLLATE setting of the location.

Declaration

Following is the declaration for strcoll() function.

int strcoll(const char *str1, const char *str2)

Parameters

- **str1** – This is the first string to be compared.
- **str2** – This is the second string to be compared.

Return Value

This functions return values that are as follows.

- if Return value < 0 then it indicates str1 is less than str2.
- if Return value > 0 then it indicates str2 is less than str1.
- if Return value = 0 then it indicates str1 is equal to str2.

Example

The following example shows the usage of strcoll() function.

```
#include <stdio.h>
#include <string.h>

int main () {
 char str1[15];
 char str2[15];
 int ret;

 strcpy(str1, "abc");
 strcpy(str2, "ABC");

 ret = strcoll(str1, str2);

 if(ret > 0) {
printf("str1 is less than str2");
 } else if(ret < 0) {
printf("str2 is less than str1");
 } else {
printf("str1 is equal to str2");
 }

 return(0);
}
```

Let us compile and run the above program that will produce the following result –

```
str1 is less than str2
```

xiii. C library function - strcpy()

Description

- The C library function **char *strcpy(char *dest, const char *src)**
- Copies the string pointed to, by src to dest.

Declaration

Following is the declaration for strcpy() function.

char *strcpy(char *dest, const char *src)

Parameters

- **dest** – This is the pointer to the destination array where the content is to be copied.
- **src** – This is the string to be copied.

Return Value

This returns a pointer to the destination string dest.

Example

The following example shows the usage of strcpy() function.

```
#include <stdio.h>
#include <string.h>

int main () {
 char src[40];
 char dest[100];

 memset(dest, '\0', sizeof(dest));
 strcpy(src, "This is studentspoint.com");
 strcpy(dest, src);

 printf("Final copied string : %s\n", dest);

 return(0);
}
```

Let us compile and run the above program that will produce the following result –

```
Final copied string : This is studentspoint.com
```

xiv. C library function - strncpy()

Description

- The C library function **char *strncpy(char *dest, const char *src, size_t n)**
- Copies up to n characters from the string pointed to, by src to dest. In a case where the length of src is less than that of n, the remainder of dest will be padded with null bytes.

Declaration

Following is the declaration for strncpy() function.

char *strncpy(char *dest, const char *src, size_t n)

Parameters

- **dest** – This is the pointer to the destination array where the content is to be copied.
- **src** – This is the string to be copied.
- **n** – The number of characters to be copied from source.

Return Value

This function returns the final copy of the copied string.

Example

The following example shows the usage of strncpy() function. Here we have used function memset() to clear the memory location.

```
#include <stdio.h>
#include <string.h>

int main () {
 char src[40];
 char dest[12];

 memset(dest, '\0', sizeof(dest));
 strcpy(src, "This is studentspoint.com");
 strncpy(dest, src, 10);

 printf("Final copied string : %s\n", dest);

 return(0);
}
```

Let us compile and run the above program that will produce the following result –

```
Final copied string : This is st
```

xv. C library function - strcspn()

Description

- The C library function size_t strcspn(const char *str1, const char *str2)
- Calculates the length of the initial segment of str1, which consists entirely of characters not in str2.

Declaration

Following is the declaration for strcspn() function.

size_t strcspn(const char *str1, const char *str2)

Parameters

- **str1** – This is the main C string to be scanned.
- **str2** – This is the string containing a list of characters to match in str1.

Return Value

This function returns the number of characters in the initial segment of string str1 which are not in the string str2.

Example

The following example shows the usage of strcspn() function.

```
#include <stdio.h>
#include <string.h>

int main () {
 int len;
 const char str1[] = "ABCDEF4960910";
 const char str2[] = "013";

 len = strcspn(str1, str2);

 printf("First matched character is at %d\n", len + 1);

 return(0);
}
```

Let us compile and run the above program that will produce the following result –

```
First matched character is at 10
```

xvi. C library function - strerror()

Description

- The C library function **char *strerror(int errnum)**
- searches an internal array for the error number errnum and returns a pointer to an error message string. The error strings produced by strerror depend on the developing platform and compiler.

Declaration

Following is the declaration for strerror() function.

char *strerror(int errnum)

Parameters

- **errnum** – This is the error number, usually errno.

Return Value

This function returns a pointer to the error string describing error errnum.

Example

The following example shows the usage of strerror() function.

```
#include <stdio.h>
#include <string.h>
#include <errno.h>

int main () {
 FILE *fp;

 fp = fopen("file.txt","r");
 if( fp == NULL ) {
printf("Error: %s\n", strerror(errno));
 }

 return(0);
}
```

Let us compile and run the above program that will produce the following result because we are trying to open a file which does not exist –

```
Error: No such file or directory
```

xvii. C library function - strlen()

Description

- The C library function **size_t strlen(const char *str)**
- Computes the length of the string str up to, but not including the terminating null character.

Declaration

Following is the declaration for strlen() function.

size_t strlen(const char *str)

Parameters

- **str** – This is the string whose length is to be found.

Return Value

This function returns the length of string.

Example

The following example shows the usage of strlen() function.

```
#include <stdio.h>
#include <string.h>

int main () {
 char str[50];
 int len;

 strcpy(str, "This is studentspoint.com");

 len = strlen(str);
 printf("Length of |%s| is |%d|\n", str, len);

 return(0);
}
```

Let us compile and run the above program that will produce the following result –

```
Length of |This is studentspoint.com| is |26|
```

xviii. C library function - strpbrk()

Description

- The C library function **char *strpbrk(const char *str1, const char *str2)**
- Finds the first character in the string str1 that matches any character specified in str2. This does not include the terminating null-characters.

Declaration

Following is the declaration for strpbrk() function.

char *strpbrk(const char *str1, const char *str2)

Parameters

- **str1** – This is the C string to be scanned.
- **str2** – This is the C string containing the characters to match.

Return Value

This function returns a pointer to the character in str1 that matches one of the characters in str2, or NULL if no such character is found.

Example

The following example shows the usage of strpbrk() function.

```
#include <stdio.h>
#include <string.h>

int main () {
 const char str1[] = "abcde2fghi3jk4l";
 const char str2[] = "34";
 char *ret;

 ret = strpbrk(str1, str2);
 if(ret) {
printf("First matching character: %c\n", *ret);
 } else {
printf("Character not found");
 }
return(0);
}
```

Let us compile and run the above program that will produce the following result –

```
First matching character: 3
```

xix. C library function - strrchr()

Description

- The C library function char *strrchr(const char *str, int c)
- Searches for the last occurrence of the character c (an unsigned char) in the string pointed to, by the argument str.

Declaration

Following is the declaration for strrchr() function.

char *strrchr(const char *str, int c)

Parameters

- **str** – This is the C string.
- **c** – This is the character to be located. It is passed as its int promotion, but it is internally converted back to char.

Return Value

This function returns a pointer to the last occurrence of character in str. If the value is not found, the function returns a null pointer.

Example

The following example shows the usage of strrchr() function.

```
#include <stdio.h>
#include <string.h>

int main () {
 int len;
 const char str[] = "http://www.studentspoint.com";
 const char ch = '.';
 char *ret;
 ret = strrchr(str, ch);

 printf("String after |%c| is - |%s|\n", ch, ret);

 return(0);
}
```

Let us compile and run the above program that will produce the following result –

```
String after |.| is - |.com|
```

xx. C library function - strspn()

Description

- The C library function **size_t strspn(const char *str1, const char *str2)**
- Calculates the length of the initial segment of str1 which consists entirely of characters in str2.

Declaration

Following is the declaration for strspn() function.

size_t strspn(const char *str1, const char *str2)

Parameters

- **str1** – This is the main C string to be scanned.
- **str2** – This is the string containing the list of characters to match in str1.

Return Value

This function returns the number of characters in the initial segment of str1 which consist only of characters from str2.

Example

The following example shows the usage of strspn() function.

```
#include <stdio.h>
#include <string.h>

int main () {
 int len;
 const char str1[] = "ABCDEFG019874";
 const char str2[] = "ABCD";

 len = strspn(str1, str2);

 printf("Length of initial segment matching %d\n", len );

 return(0);
}
```

Let us compile and run the above program that will produce the following result –

```
Length of initial segment matching 4
```

xxi. C library function - strstr()

Description

- The C library function **char *strstr(const char *haystack, const char *needle)**
- Function finds the first occurrence of the substring needle in the string haystack. The terminating '\0' characters are not compared.

Declaration

Following is the declaration for strstr() function.

char *strstr(const char *haystack, const char *needle)

Parameters

- **haystack** – This is the main C string to be scanned.
- **needle** – This is the small string to be searched with-in haystack string.

Return Value

This function returns a pointer to the first occurrence in haystack of any of the entire sequence of characters specified in needle, or a null pointer if the sequence is not present in haystack.

Example

The following example shows the usage of strstr() function.

```
#include <stdio.h>
#include <string.h>

int main () {
 const char haystack[20] = "StudentsPoint";
 const char needle[10] = "Point";
 char *ret;

 ret = strstr(haystack, needle);

```

```
printf("The substring is: %s\n", ret);

return(0);
}
```

Let us compile and run the above program that will produce the following result −

```
The substring is: Point
```

xxii. C library function - strtok()

Description

- The C library function **char *strtok(char *str, const char *delim)**
- Breaks string str into a series of tokens using the delimiter delim.

Declaration

Following is the declaration for strtok() function.

char *strtok(char *str, const char *delim)

Parameters

- **str** − The contents of this string are modified and broken into smaller strings (tokens).
- **delim** − This is the C string containing the delimiters. These may vary from one call to another.

Return Value

This function returns a pointer to the last token found in the string. A null pointer is returned if there are no tokens left to retrieve.

Example

The following example shows the usage of strtok() function.

```
#include <string.h>
#include <stdio.h>

int main () {
 char str[80] = "This is - www. studentspoint.com - website";
 const char s[2] = "-";
```

```
char *token;

/* get the first token */
token = strtok(str, s);

/* walk through other tokens */
while( token != NULL ) {
printf( " %s\n", token );

token = strtok(NULL, s);
}

return(0);
}
```

Let us compile and run the above program that will produce the following result – *This*

```
is
www.studentspoint.com
website
```

xxiii. C library function - strxfrm()

Description

- The C library function **size_t strxfrm(char *dest, const char *src, size_t n)**
- Transforms the first n characters of the string src into current locale and place them in the string dest.

Declaration

Following is the declaration for strxfrm() function.

size_t strxfrm(char *dest, const char *src, size_t n)

Parameters

- **dest** – This is the pointer to the destination array where the content is to be copied. It can be a null pointer if the argument for n is zero.

- **src** – This is the C string to be transformed into current locale.
- **n** – The maximum number of characters to be copied to str1.

Return Value

This function returns the length of the transformed string, not including the terminating null-character.

Example

The following example shows the usage of strxfrm() function.

```
#include <stdio.h>
#include <string.h>

int main () {
 char dest[20];
 char src[20];
 int len;

 strcpy(src, " Students Point");
 len = strxfrm(dest, src, 20);

 printf("Length of string |%s| is: |%d|", dest, len);

 return(0);
}
```

Let us compile and run the above program that will produce the following result –

```
Length of string |Students Point| is: |15|
```

3.4. Built in Function: C Library - <math.h>

C Library math.h functions

The **math.h** header defines various mathematical functions and one macro. All the functions available in this library take **double** as an argument and return **double** as the result. Let us discuss some important functions one by one.

Library Macros

There is only one macro defined in this library.

HUGE_VAL - This macro is used when the result of a function may not be representable as a floating point number. If magnitude of the correct result is too large to be represented, the function sets errno to ERANGE to indicate a range error, and returns a particular, very large value named by the macro HUGE_VAL or its negation (- HUGE_VAL).

If the magnitude of the result is too small, a value of zero is returned instead. In this case, errno might or might not be set to ERANGE.

i. **double ceil(double x):** The C library function double ceil(double x) returns the smallest integer value greater than or equal to x.

Syntax: *double ceil(double x)*

```
// C code to illustrate
// the use of ceil function.
#include <stdio.h>
#include <math.h>

int main ()
{
float val1, val2, val3, val4;

val1 = 1.6;
val2 = 1.2;
val3 = -2.8;
val4 = -2.3;

printf ("value1 = %.1lf\n", ceil(val1));
printf ("value2 = %.1lf\n", ceil(val2));
printf ("value3 = %.1lf\n", ceil(val3));
printf ("value4 = %.1lf\n", ceil(val4));

return(0);
}
```

Output

```
value1 = 2.0
value2 = 2.0
```

```
value3 = -2.0
value4 = -2.0
```

ii. **double floor(double x):** The C library function double floor(double x) returns the largest integer value less than or equal to x.

Syntax: *double floor(double x)*

```
// C code to illustrate
// the use of floor function
#include <stdio.h>
#include <math.h>

int main ()
{
float val1, val2, val3, val4;

val1 = 1.6;
val2 = 1.2;
val3 = -2.8;
val4 = -2.3;

printf("Value1 = %.1lf\n", floor(val1));
printf("Value2 = %.1lf\n", floor(val2));
printf("Value3 = %.1lf\n", floor(val3));
printf("Value4 = %.1lf\n", floor(val4));

return(0);
}
```

Output

```
Value1 = 1.0
Value2 = 1.0
Value3 = -3.0
Value4 = -3.0
```

iii. **double fabs(double x):** The C library function double fabs(double x) returns the absolute value of x.

Syntax: *double fabs(double x)*

```
// C code to illustrate
// the use of fabs function
#include <stdio.h>
#include <math.h>

int main ()
{
 int a, b;
 a = 1234;
 b = -344;

 printf("The absolute value of %d is %lf\n", a, fabs(a));
 printf("The absolute value of %d is %lf\n", b, fabs(b));

 return(0);
}
```

Output

```
The absolute value of 1234 is 1234.000000
The absolute value of -344 is 344.000000
```

iv. **double log(double x):** The C library function double log(double x) returns the natural logarithm (base-e logarithm) of x.

Syntax: *double log(double x)*

```
// C code to illustrate
// the use of log function

#include <stdio.h>
#include <math.h>

int main ()
{
 double x, ret;
 x = 2.7;

 /* finding log(2.7) */
 ret = log(x);
```

```
printf("log(%lf) = %lf", x, ret);

return(0);
}
```

Output:

```
log(2.700000) = 0.993252
```

v. **double log10(double x):** The C library function double log10(double x) returns the common logarithm (base-10 logarithm) of x.

Syntax: double log10(double x)

```
// C code to illustrate
// the use of log10 function
#include <stdio.h>
#include <math.h>

int main ()
{
 double x, ret;
 x = 10000;

 /* finding value of log1010000 */
 ret = log10(x);
 printf("log10(%lf) = %lf\n", x, ret);

 return(0);
}
```

Output

```
log10(10000.000000) = 4.000000
```

vi. **double fmod(double x, double y):** The C library function double fmod(double x, double y) returns the remainder of x divided by y.

Syntax: double fmod(double x, double y)

```
// C code to illustrate
// the use of fmod function
#include <stdio.h>
#include <math.h>
```

```
int main ()
{
 float a, b;
 int c;
 a = 8.2;
 b = 5.7;
 c = 3;
 printf("Remainder of %f / %d is %lf\n", a, c, fmod(a, c));
 printf("Remainder of %f / %f is %lf\n", a, b, fmod(a, b));

 return(0);
}
```

Output

```
Remainder of 8.200000 / 3 is 2.200000
Remainder of 8.200000 / 5.700000 is 2.500000
```

vii. **double sqrt(double x):** The C library function double sqrt(double x) returns the square root of x.

Syntax: *double sqrt(double x)*

```
// C code to illustrate
// the use of sqrt function
#include <stdio.h>
#include <math.h>

int main ()
{

 printf("Square root of %lf is %lf\n", 225.0, sqrt(225.0) );
 printf("Square root of %lf is %lf\n", 300.0, sqrt(300.0) );

 return(0);
}\
```

Output

```
Square root of 225.000000 is 15.000000
Square root of 300.000000 is 17.320508
```

viii. **double pow(double x, double y):** The C library function double pow(double x, double y) returns x raised to the power of y i.e. xy.

Syntax: double pow(double x, double y)

```
// C code to illustrate
// the use of pow function
#include <stdio.h>
#include <math.h>

int main ()
{
 printf("Value 8.0 ^ 3 = %lf\n", pow(8.0, 3));

 printf("Value 3.05 ^ 1.98 = %lf", pow(3.05, 1.98));

 return(0);
}
```

Output:

```
Value 8.0 ^ 3 = 512.000000
Value 3.05 ^ 1.98 = 9.097324
```

ix. **double modf(double x, double *integer):** The C library function double modf(double x, double *integer) returns the fraction component (part after the decimal), and sets integer to the integer component.

Syntax: double modf(double x, double *integer)

```
// C code to illustrate
// the use of modf function
#include<stdio.h>
#include<math.h>

int main ()
{
 double x, fractpart, intpart;

 x = 8.123456;
fractpart = modf(x, &intpart);
```

```
printf("Integral part = %lf\n", intpart);
printf("Fraction Part = %lf \n", fractpart);

return(0);
}
```

Output

```
Integral part = 8.000000
Fraction Part = 0.123456
```

x. **double exp(double x):** The C library function double exp(double x) returns the value of e raised to the xth power.

Syntax: double exp(double x)

Following code represents

```
// C code to illustrate
// the use of exp function
#include <stdio.h>
#include <math.h>

int main ()
{
 double x = 0;

 printf("The exponential value of %lf is %lf\n", x, exp(x));
 printf("The exponential value of %lf is %lf\n", x+1, exp(x+1));
 printf("The exponential value of %lf is %lf\n", x+2, exp(x+2));

 return(0);
}
```

Output

```
The exponential value of 0.000000 is 1.000000
The exponential value of 1.000000 is 2.718282
The exponential value of 2.000000 is 7.389056
```

xi. **double cos(double x):** The C library function double cos(double x) returns the cosine of a radian angle x.

Syntax: *double cos(double x)*

Note: Same syntax can be used for other trignometric functions like sin, tan etc.

```
// C code to illustrate
// the use of cos function
#include <stdio.h>
#include <math.h>

#define PI 3.14159265

int main ()
{
 double x, ret, val;

 x = 60.0;
 val = PI / 180.0;
 ret = cos( x*val );
 printf("The cosine of %lf is %lf degrees\n", x, ret);

 x = 90.0;
 val = PI / 180.0;
 ret = cos( x*val );
 printf("The cosine of %lf is %lf degrees\n", x, ret);

 return(0);
}
```

Output

```
The cosine of 60.000000 is 0.500000 degrees
The cosine of 90.000000 is 0.000000 degrees
```

xii. double acos(double x): The C library function double acos(double x) returns the arc cosine of x in radians.

Syntax: *double acos(double x)*

Note: Same syntax can be used for other arc trignometric functions like asin, atan etc.

```
// C code to illustrate
// the use of acos function
#include <stdio.h>
```

```
#include <math.h>

#define PI 3.14159265

int main ()
{
 double x, ret, val;

 x = 0.9;
 val = 180.0 / PI;

 ret = acos(x) * val;
 printf("The arc cosine of %lf is %lf degrees", x, ret);

 return(0);
}
```

Output

```
The arc cosine of 0.900000 is 25.855040 degrees
```

xiii. double tanh(double x): The C library function double tanh(double x) returns the hyperbolic tangent of x.

Syntax: *double tanh(double x)*

Note: Same syntax can be used for other hyperbolic trignometric functions like sinh, cosh etc.

```
// C code to illustrate
// the use of tanh function
#include <stdio.h>
#include <math.h>

int main ()
{
 double x, ret;
 x = 0.5;

 ret = tanh(x);
 printf("The hyperbolic tangent of %lf is %lf degrees", x, ret);

```

```
return(0);
}
```

Output

```
The hyperbolic tangent of 0.500000 is 0.462117 degrees
```

3.5. Recursion in C

Definition

Recursion is the process of repeating items in a self-similar way. In programming languages, if a program allows you to call a function inside the same function, then it is called a recursive call of the function.

```
void recursion() {
 recursion(); /* function calls itself */
}

int main() {
 recursion();
}
```

- The C programming language supports recursion, i.e., a function to call itself. But while using recursion, programmers need to be careful to define an exit condition from the function, otherwise it will go into an infinite loop.
- Recursive functions are very useful to solve many mathematical problems, such as calculating the factorial of a number, generating Fibonacci series, etc.

How recursion works?

```
void recurse()
{
... .. ...
recurse();
... .. ...
}

int main()
{
... .. ...
```

```
recurse();
... .. ...
}
```

Number Factorial

The following example calculates the factorial of a given number using a recursive function.

```
#include <stdio.h>

int factorial(unsigned int i) {

 if(i <= 1) {
return 1;
 }
 return i * factorial(i - 1);
}

intmain() {
 int i = 12;
 printf("Factorial of %d is %d\n", i, factorial(i));
 return 0;
}
```

When the above code is compiled and executed, it produces the following result.

```
Factorial of 12 is 479001600
```

Fibonacci Series

The following example generates the Fibonacci series for a given number using a recursive function

```
#include <stdio.h>

int fibonacci(int i) {

 if(i == 0) {
return 0;
 }
```

```
if(i == 1) {
return 1;
}
return fibonacci(i-1) + fibonacci(i-2);
}

intmain() {

int i;

for (i = 0; i < 10; i++) {
printf("%d\t\n", fibonacci(i));
}

return 0;
}
```

When the above code is compiled and executed, it produces the following result

```
0
1
1
2
3
5
8
13
21
34
```

Example of Sum of Natural Numbers Using Recursion

```
#include <stdio.h>
int sum(int n);

int main()
{
int number, result;
```

```
printf("Enter a positive integer: ");
scanf("%d", &number);

result = sum(number);

printf("sum=%d", result);
}

int sum(int num)
{
if (num!=0)
return num + sum(num-1); // sum() function calls itself
else
return num;
}
```

Output

```
Enter a positive integer:
3
6
```

Advantages and Disadvantages of Recursion

- Recursion makes program elegant and cleaner. All algorithms can be defined recursively which makes it easier to visualize and prove.
- If the speed of the program is vital then, you should avoid using recursion. Recursions use more memory and are generally slow. Instead, you can use loop.

3.6. Example Program

3.6.1. Example program of computation of sine series using built in function

C program for Sine Series

Before going to the program for Sine Series first let us understand what is a Sine Series?

Sine Series

Sine Series is a series which is used to find the value of Sin(x). where, x is the angle in degree which is converted to Radian. The formula used to express the Sin(x) as Sine Series is

$$\sin x = \sum_{x=0}^{\infty} (-1)^n \frac{x^{2n+1}}{(2n+1)!}$$

Expanding the above notation, the formula of Sine Series is

$$\sin x = x - \frac{x^3}{3!} + \frac{x^5}{5!} - \frac{x^7}{7!} + \ldots$$

For example,

Let the value of x be 30.

$$x = 30 * \frac{\pi}{180} = 30 * \frac{3.14159}{180} = 0.52359$$

So, Radian value for 30 degree is 0.52359.

$$\text{Sin}(0.52359) = 0.52359 - \frac{0.52359^3}{3!} + \frac{0.52359^5}{5!} - \frac{0.52359^7}{7!} + \ldots\ldots\ldots$$

So, the value of Sin(30) is 0.5.

Program

```
#include<stdio.h>
#include<conio.h>

void main()
{
int i, n;
float x, sum, t;
clrscr();

printf(" Enter the value for x : ");
scanf("%f",&x);

printf(" Enter the value for n : ");
scanf("%d",&n);
```

```
x=x*3.14159/180;
t=x;
sum=x;

/* Loop to calculate the value of Sine */
for(i=1;i<=n;i++)
{
t=(t*(-1)*x*x)/(2*i*(2*i+1));
sum=sum+t;
}

printf(" The value of Sin(%f) = %.4f",x,sum);
getch();
}
```

Output

```
Enter the value for x: 45
Enter the vale for n: 5
The value of sin(0.785398) = 0.7071
```

Step by Step working of the above Program Code:

Let us assume that the user enters the value of _x' as **45** and _n' as **4.**

- Converting _x' to radian value

x = x * 3.14159 / 180(x = 45 * 3.14159 / 180)So,**x=0.785398**

- It assigns t=x and sum=x (i.e. **t=0.785398** and **sum=0.785398**)
- It assigns the value of **i=1** and the loop continues till the condition of the for loop is true.
- i<=n(**1<=4**)for loop condition is true
  t = (0.785398 * (-1) * 0.785398 * 0.785398)/(2 * 1 * (2 * 1 + 1))
  So,**t = - 0.08074**
  sum = 0.785398 + (- 0.08074)
  So,**sum=0.70465**
  i++
  So,**i=2**
- i<=n(**2<=4**)for loop condition is true
  t = (- 0.08074 * (-1) * 0.785398 * 0.785398)/(2 * 2 * (2 * 2 + 1))
  So,**t = 0.00249**
  sum = 0.70465 + 0.00249
  So,**sum=0.70714**

i++
So,**i=3**

- i<=n(**3<=4**)for loop condition is true
  t = (0.00249 * (-1) * 0.785398 * 0.785398)/(2 * 3 * (2 * 3 + 1))
  So,**t = - 0.000032**
  sum = 0.70714 + (- 0.000032)
  So,**sum=0.707108**
  i++
  So,**i=4**
- i<=n(**4<=4**)for loop condition is true
  t = (- 0.000032 * (-1) * 0.785398 * 0.785398)/(2 * 4 * (2 * 4 + 1))
  So,**t = 0.000000274**
  sum = 0.707108 + 0.000000274
  So,**sum=0.707108274**
  i++
  So,**i=5**
- i<=n(**5<=4**)for loop condition is false
  It comes out of the for loop.
- Finally it prints **The value ofSin(0.785398) = 0.7071**
- Thus program execution is completed.

3.6.2. Example program of scientific calculator using built in function

A sample C programming code for a real time calculator application program is given below. This program will perform the below calculator operations.

- Addition
- Subtraction
- Multiplication
- Division
- Modulus
- Power
- Factorial

In below c programming code example, the user is prompted to choose the operations(i.e. addition, subtraction etc)to be performed and then prompted to key in the values which are used to perform the operations. Then, the result will be shown as output to the user.

Program

```
// Calculator example using C code
#include<stdio.h>
#include<conio.h>
#include<math.h>
#include<stdlib.h>

#define KEY "Enter the calculator Operation you want to do:"

// Function prototype declaration
void addition();
void subtraction();
void multiplication();
void division();
void modulus();
void power();
int factorial();
void calculator_operations();

// Start of Main Program
int main()
{
int X=1;
char Calc_oprn;

// Function call
calculator_operations();

while(X)
{ printf("\n");
printf("%s : ", KEY);

Calc_oprn=getche();

switch(Calc_oprn)
{
```

```
case '+': addition();
break;

case '-': subtraction();
break;

case '*': multiplication();
break;

case '/': division();
break;

case '?': modulus();
break;

case '!': factorial();
break;

case '^': power();
break;

case 'H':
case 'h': calculator_operations();
break;

case 'Q':
case 'q': exit(0);
break;
case 'c':
case 'C': system("cls");
calculator_operations();
break;

default : system("cls");

printf("\n**********You have entered unavailable option");
printf("***********\n");
printf("\n*****Please Enter any one of below available ");
```

```
printf("options****\n");
calculator_operations();
}
}
}

//Function Definitions

void calculator_operations()
{
//system("cls");use system function to clear
//screen instead of clrscr();
printf("\n Welcome to C calculator \n\n");

printf("******* Press 'Q' or 'q' to quit ");
printf("the program ********\n");
printf("***** Press 'H' or 'h' to display ");
printf("below options *****\n\n");
printf("Enter 'C' or 'c' to clear the screen and");
printf(" display available option \n\n");

printf("Enter + symbol for Addition \n");
printf("Enter - symbol for Subtraction \n");
printf("Enter * symbol for Multiplication \n");
printf("Enter / symbol for Division \n");
printf("Enter ? symbol for Modulus\n");
printf("Enter ^ symbol for Power \n");
printf("Enter ! symbol for Factorial \n\n");
}

void addition()
{
int n, total=0, k=0, number;
printf("\nEnter the number of elements you want to add:");
scanf("%d",&n);
printf("Please enter %d numbers one by one: \n",n);
while(k<n)
```

```
{
scanf("%d",&number);
total=total+number;
k=k+1;
}
printf("Sum of %d numbers = %d \n",n,total);
}

void subtraction()
{
int a, b, c = 0;
printf("\nPlease enter first number: ");
scanf("%d", &a);
printf("Please enter second number : ");
scanf("%d", &b);
c = a - b;
printf("\n%d - %d = %d\n", a, b, c);
}

void multiplication()
{
int a, b, mul=0;
printf("\nPlease enter first numb : ");
scanf("%d", &a);
printf("Please enter second number: ");
scanf("%d", &b);
mul=a*b;
printf("\nMultiplication of entered numbers = %d\n",mul);
}

void division()
{
int a, b, d=0;
printf("\nPlease enter first number: ");
scanf("%d", &a);
printf("Please enter second number : ");
scanf("%d", &b);
```

```
d=a/b;
printf("\nDivision of entered numbers=%d\n",d);
}

void modulus()
{
int a, b, d=0;
printf("\nPlease enter first number : ");
scanf("%d", &a);
printf("Please enter second number: ");
scanf("%d", &b);
d=a%b;
printf("\nModulus of entered numbers = %d\n",d);
}

void power()
{
double a,num, p;
printf("\nEnter two numbers to find the power \n");
printf("number: ");
scanf("%lf",&a);

printf("power : ");
scanf("%lf",&num);

p=pow(a,num);

printf("\n%lf to the power %lf = %lf \n",a,num,p);
}

int factorial()
{
int i,fact=1,num;

printf("\nEnter a number to find factorial : ");
scanf("%d",&num);
```

```
if (num<0)
{
printf("\nPlease enter a positive number to");
printf(" find factorial and try again. \n");
printf("\nFactorial can't be found for negative");
printf(" values. It can be only positive or 0\n");
return 1;
}

for(i=1;i<=num;i++)
fact=fact*i;
printf("\n");
printf("Factorial of entered number %d is:%d\n",num,fact);
return 0;
}
```

Output

```
            Welcome to C calculator

******* Press 'Q' or 'q' to quit the program ********
***** Press 'H' or 'h' to display below options *****

Enter 'C' or 'c' to clear the screen and display available option

Enter + symbol for Addition
Enter - symbol for Subtraction
Enter * symbol for Multiplication
Enter / symbol for Division
Enter ? symbol for Modulus
Enter ^ symbol for Power
Enter ! symbol for Factorial


Enter the calculator Operation you want to do: : +
Enter the number of elements you want to add : 3
Please enter 3 numbers one by one:
10
20
30
Sum of 3 numbers = 60

Enter the calculator Operation you want to do: : -
Please enter first number  : 40
Please enter second number : 25

40 - 25 = 15

Enter the calculator Operation you want to do: : *
Please enter first numb   : 3
Please enter second number: 6

Multiplication of entered numbers = 18

Enter the calculator Operation you want to do: : /
Please enter first number  : 30
Please enter second number : 11

Division of entered numbers=2
```

```
Enter the calculator Operation you want to do: : ?
Please enter first number   : 30
Please enter second number  : 11

Modulus of entered numbers = 8

Enter the calculator Operation you want to do: : ^
Enter two numbers to find the power
number: 3
power : 4

3.000000 to the power 4.000000 = 81.000000

Enter the calculator Operation you want to do: : !
Enter a number to find factorial : 4

Factorial of entered number 4 is : 24
```

3.7. Binary Search using Recursive Functions in C

Write a simple code for binary search using function recursion in c programming language.

```
#include<stdio.h>
int main(){

int a[10],i,n,m,c,l,u;

printf("Enter the size of an array: ");
scanf("%d",&n);

printf("Enter the elements of the array: " );
for(i=0;i<n;i++){
 scanf("%d",&a[i]);
}

printf("Enter the number to be search: ");
scanf("%d",&m);

l=0,u=n-1;
c=binary(a,n,m,l,u);
if(c==0)
 printf("Number is not found.");
else
 printf("Number is found.");
```

```
return 0;
 }

int binary(int a[],int n,int m,int l,int u){

 int mid,c=0;

 if(l<=u){
mid=(l+u)/2;
if(m==a[mid]){
c=1;
}
else if(m<a[mid]){
return binary(a,n,m,l,mid-1);
}
else
return binary(a,n,m,mid+1,u);
 } else
 return c;
}
```

Output

```
Enter the size of an array: 5
Enter the elements of the array: 8 9 10 11 12
Enter the number to be search: 8
Number is found.
```

3.8. Pointers in C

- Pointers in C are easy and fun to learn.
- Some C programming tasks are performed more easily with pointers, and other tasks, such as dynamic memory allocation, cannot be performed without using pointers.
- So it becomes necessary to learn pointers to become a perfect C programmer. Let's start learning them in simple and easy steps.
- As you know, every variable is a memory location and every memory location has its address defined which can be

accessed using ampersand (&) operator, which denotes an address in memory.

Consider the following example, which prints the address of the variables defined.

```
#include <stdio.h>

int main () {

 intvar1;
 char var2[10];

 printf("Address of var1 variable: %x\n", &var1);
 printf("Address of var2 variable: %x\n", &var2);

 return 0;
}
```

When the above code is compiled and executed, it produces the following result.

```
Address of var1 variable: bff5a400
Address of var2 variable: bff5a3f6
```

What are Pointers?

A pointer is a variable whose value is the address of another variable, i.e., direct address of the memory location. Like any variable or constant, you must declare a pointer before using it to store any variable address. The general form of a pointer variable declaration is –

*type *var-name;*

Here, type is the pointer's base type; it must be a valid C data type and **var-name** is the name of the pointer variable. The asterisk * used to declare a pointer is the same asterisk used for multiplication. However, in this statement the asterisk is being used to designate a variable as a pointer. Take a look at some of the valid pointer declarations.

```
int*ip;/* pointer to an integer */
double *dp;/* pointer to a double */
float*fp;/* pointer to a float */
char *ch /* pointer to a character */
```

The actual data type of the value of all pointers, whether integer, float, character, or otherwise, is the same, a long hexadecimal number that represents a memory address. The only difference between pointers of different data types is the data type of the variable or constant that the pointer points to.

How to Use Pointers?

There are a few important operations, which we will do with the help of pointers very frequently

- We define a pointer variable,
- Assign the address of a variable to a pointer and
- Finally access the value at the address available in the pointer variable. This is done by using unary operator * that returns the value of the variable located at the address specified by its operand. The following example makes use of these operations

```
#include <stdio.h>

int main () {

intvar = 20; /* actual variable declaration */
int*ip;/* pointer variable declaration */

ip = &var;/* store address of var in pointer variable*/

printf("Address of var variable: %x\n", &var);

/* address stored in pointer variable */
printf("Address stored in ip variable: %x\n", ip );

/* access the value using the pointer */
printf(" Value of *ip variable: %d\n", *ip );

return 0;
}
```

Output

```
Address of var variable: bffd8b3c
Address stored in ip variable: bffd8b3c
Value of *ip variable: 20
```

NULL Pointers

- It is always a good practice to assign a NULL value to a pointer variable in case you do not have an exact address to be assigned. This is done at the time of variable declaration. A pointer that is assigned NULL is called a **null** pointer.
- The NULL pointer is a constant with a value of zero defined in several standard libraries. Consider the following program

```
#include <stdio.h>

int main () {

 int*ptr = NULL;

 printf("The value of ptr is : %x\n", ptr);

 return 0;
}
```

Output

```
The value of ptr is 0
```

- In most of the operating systems, programs are not permitted to access memory at address 0 because that memory is reserved by the operating system.
- However, the memory address 0 has special significance; it signals that the pointer is not intended to point to an accessible memory location.
- But by convention, if a pointer contains the null (zero) value, it is assumed to point to nothing.

To check for a null pointer, you can use an 'if' statement as follows.

```
if(ptr) /* succeeds if p is not null */
if(!ptr)/* succeeds if p is null */
```

Pointers in Detail

- Pointers have many but easy concepts and they are very important to C programming. The following important pointer concepts should be clear to any C programmer.

3.9. Pointer Arithmetic

- A pointer in c is an address, which is a numeric value. Therefore, you can perform arithmetic operations on a pointer just as you can on a numeric value. There are four arithmetic operators that can be used on pointers: ++, --, +, and -.
- To understand pointer arithmetic, let us consider that ptr is an integer pointer which points to the address 1000. Assuming 32-bit integers, let us perform the following arithmetic operation on the pointer.

```
ptr++
```

After the above operation, the ptr will point to the location 1004 because each time ptr is incremented, it will point to the next integer location which is 4 bytes next to the current location. This operation will move the pointer to the next memory location without impacting the actual value at the memory location. If ptr points to a character whose address is 1000, then the above operation will point to the location 1001 because the next character will be available at 1001.

Incrementing a Pointer

We prefer using a pointer in our program instead of an array because the variable pointer can be incremented, unlike the array name which cannot be incremented because it is a constant pointer.

The following program increments the variable pointer to access each succeeding element of the array

```
#include <stdio.h>

const int MAX = 3;

int main () {
```

```
intvar[] = {10, 100, 200};
inti, *ptr;

/* let us have array address in pointer */
ptr = var;

for ( i = 0; i < MAX; i++) {

printf("Address of var[%d] = %x\n", i, ptr );
printf("Value of var[%d] = %d\n", i, *ptr );

/* move to the next location */
ptr++;
 }

 return 0;
}
```

Output

```
Address of var[0] = bf882b30
Value of var[0] = 10
Address of var[1] = bf882b34
Value of var[1] = 100
Address of var[2] = bf882b38
Value of var[2] = 200
```

Decrementing a Pointer

The same considerations apply to decrementing a pointer, which decreases its value by the number of bytes of its data type as shown below

```
#include <stdio.h>

const int MAX = 3;

int main () {

 intvar[] = {10, 100, 200};
 inti, *ptr;

 /* let us have array address in pointer */
 ptr = &var[MAX-1];
```

```
 for ( i = MAX; i > 0; i--) {

printf("Address of var[%d] = %x\n", i-1, ptr );
printf("Value of var[%d] = %d\n", i-1, *ptr );

/* move to the previous location */
ptr--;
 }

 return 0;
}
```

Output

```
Address of var[2] = bfedbcd8
Value of var[2] = 200
Address of var[1] = bfedbcd4
Value of var[1] = 100
Address of var[0] = bfedbcd0
Value of var[0] = 10
```

Pointer Comparisons

Pointers may be compared by using relational operators, such as ==, <, and >. If p1 and p2 point to variables that are related to each other, such as elements of the same array, then p1 and p2 can be meaningfully compared.

The following program modifies the previous example – one by incrementing the variable pointer so long as the address to which it points is either less than or equal to the address of the last element of the array, which is &var[MAX - 1].

```
#include   <stdio.h>

const int MAX = 3;

int main () {

 intvar[] = {10, 100, 200};
 inti, *ptr;

/* let us have address of the first element in pointer */
ptr = var;
i = 0;
```

```
while ( ptr <= &var[MAX - 1] ) {

printf("Address of var[%d] = %x\n", i, ptr );
printf("Value of var[%d] = %d\n", i, *ptr );

/* point to the previous location */
ptr++;
i++;
}

return 0;
}
```

Output

```
Address of var[0] = bfdbcb20
Value of var[0] = 10
Address of var[1] = bfdbcb24
Value of var[1] = 100
Address of var[2] = bfdbcb28
Value of var[2] = 200
```

3.10. Arrays and Pointers

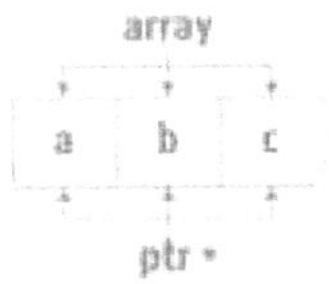

Figure: arrays and pointers

Arrays are closely related to pointers in C programming but the important difference between them is that, a pointer variable takes different addresses as value whereas, in case of array it is fixed.

Example:

```
#include <stdio.h>
int main()
{
 char charArr[4];
 int i;

 for(i = 0; i < 4; ++i)
 {
printf("Address of charArr[%d] = %u\n", i, &charArr[i]);
```

```
}

return 0;
}
```

Output

```
Address of charArr[0] = 28ff44
Address of charArr[1] = 28ff45
Address of charArr[2] = 28ff46
Address of charArr[3] = 28ff47
```

Note: You may get different address of an array.

- Notice, that there is an equal difference (difference of 1 byte) between any two consecutive elements of array charArr.
- But, since pointers just point at the location of another variable, it can store any address.

Relation between Arrays and Pointers

Consider an array: *int arr[4];*

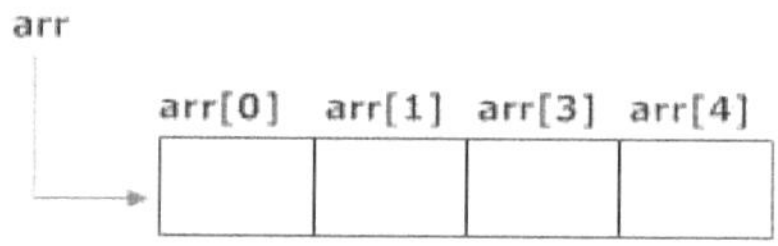

Figure: Array as Pointer

- In C programming, name of the array always points to address of the first element of an array.
- In the above example, arr and &arr[0] points to the address of the first element.

&arr[0] is equivalent to arr

Since, the addresses of both are the same, the values of arr and &arr[0] are also the same.

*arr[0] is equivalent to *arr (value of an address of the pointer)*

Similarly,

```
&arr[1] is equivalent to (arr + 1) AND, arr[1] is equivalent to *(arr + 1).
&arr[2] is equivalent to (arr + 2) AND, arr[2] is equivalent to *(arr + 2).
&arr[3] is equivalent to (arr + 3) AND, arr[3] is equivalent to *(arr + 3).
```

*&arr[i] is equivalent to (arr + i) AND, arr[i] is equivalent to *(arr + i).*

In C, you can declare an array and can use pointer to alter the data of an array.

Example of Program to find the sum of six numbers with arrays and pointers

```
#include <stdio.h>
int main()
{
int i, classes[6],sum = 0;
printf("Enter 6 numbers:\n");
for(i = 0; i < 6; ++i)
{
// (classes + i) is equivalent to &classes[i]
scanf("%d",(classes + i));

// *(classes + i) is equivalent to classes[i]
sum += *(classes + i);
}
printf("Sum = %d", sum);
return 0;
}
```

Output

```
Enter 6 numbers:
2
3
4
5
3
4
Sum = 21
```

3.11. Array of Pointers in C

Before we understand the concept of arrays of pointers, let us consider the following example, which uses an array of 3 integers.

```
#include <stdio.h>


const int MAX = 3;


int main () {
```

```
intvar[] = {10, 100, 200};
int i;

for (i = 0; i < MAX; i++) {
printf("Value of var[%d] = %d\n", i, var[i] );
}

return 0;
}
```

Output

```
Value of var[0] = 10
Value of var[1] = 100
Value of var[2] = 200
```

There may be a situation when we want to maintain an array, which can store pointers to an **int** or **char** or any other data type available. Following is the declaration of an array of pointers to an integer.

```
int *ptr[MAX];
```

It declares ptr as an array of MAX integer pointers. Thus, each element in ptr, holds a pointer to an int value. The following example uses three integers, which are stored in an array of pointers, as follows

```
#include <stdio.h>

const int MAX = 3;

int main () {

intvar[] = {10, 100, 200};
int i, *ptr[MAX];

for ( i = 0; i < MAX; i++) {
ptr[i] = &var[i]; /* assign the address of integer. */
}
```

```
 for ( i = 0; i < MAX; i++) {
printf("Value of var[%d] = %d\n", i, *ptr[i] );
 }

 return 0;
}\
```

Output

```
Value of var[0] = 10
Value of var[1] = 100
Value of var[2] = 200
```

You can also use an array of pointers to character to store a list of strings as follows.

```
#include <stdio.h>

const int MAX = 4;

int main () {

 char *names[] = {
"Zara Ali",
"Hina  Ali",
"Nuha  Ali",
"Sara Ali"
 };

 int i = 0;
 for ( i = 0; i < MAX; i++) {
printf("Value of names[%d] = %s\n", i, names[i] );
 }

 return 0;
}
```

Output

```
Value of names[0] = Zara Ali
Value of names[1] = Hina Ali
Value of names[2] = Nuha Ali
Value of names[3] = Sara Ali
```

3.12. Example Program: Sorting of names

Write a program to sort a list of strings in alphabetical order, using an array of pointers.

```
#include<stdio.h>
void main()
{
char *T;
int I,J,K;
char *ARRAY[5]={"SUNIL","ANIL","DILIP","JAY","BHARAT"};
clrscr();
for(I=0;I<5;I++)
{
 printf("%s \t",ARRAY[I]);
} printf("\n");
for(I=0;I<4;I++)
{
for(J=0;J<4-I;J++)
{
 K=strcmp(ARRAY[J],ARRAY[J+1]);
 if(K>0)
 { T=ARRAY[J];
 ARRAY[J]=ARRAY[J+1];
 ARRAY[J+1]=T;
 }
}
}
for(I=0;I<5;I++)
{
 printf("%s \t",ARRAY[I]);
}
getch();
}
```

Output

```
SUNIL  ANIL          DILIP         JAY     BHARAT

ANIL         BHARAT        DILIP         JAY     SUNIL
```

3.13. Parameter passing in C

Passing Argument to Function:

- In C Programming we have different ways of parameter passing schemes such as Call by Value and Call by Reference.
- Function is good programming style in which we can write reusable code that can be called whenever require.
- Whenever we call a function then sequence of executable statements gets executed. We can pass some of the information to the function for processing called argument.

Two Ways of Passing Argument to Function in C Language:

- Call by Reference
- Call by Value

Let us discuss different ways one by one

3.13.1. Pass by value

Example

```
#include<stdio.h>

void interchange(int number1,int number2)
{
int temp;
temp = number1;
number1 = number2;
number2 = temp;
}

int main() {

int num1=50,num2=70;
interchange(num1,num2);
```

```
printf(" \nNumber 1 : %d",num1);
printf(" \nNumber 2 : %d",num2);

return(0);
}
```

Output

Number 1 : 50

Number 2 : 70

Explanation : Call by Value

- While Passing Parameters using call by value , xerox copy of original parameter is created and passed to the called function.
- Any update made inside method will not affect the original value of variable in calling function.
- In the above example num1 and num2 are the original values and xerox copy of these values is passed to the function and these values are copied into number1,number2 variable of sum function respectively.
- As their scope is limited to only function so they cannot alter the values inside main function.

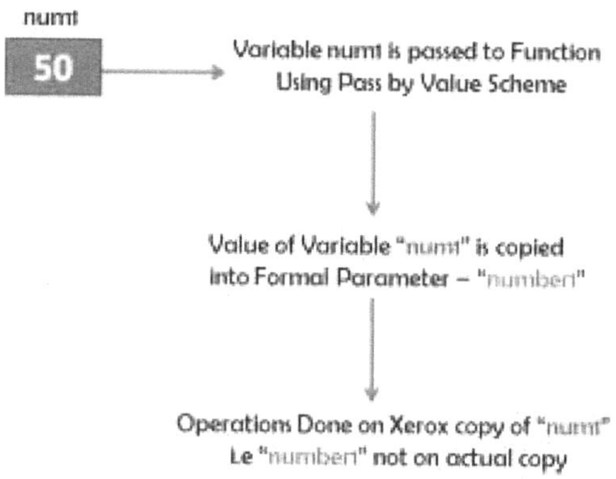

Figure: Pass by value

3.13.2. PASS BY REFERENCE/POINTER/ADDRESS

```
#include<stdio.h>

void interchange(int *num1,int *num2)
{
int temp;
temp= *num1;
*num1 = *num2;
*num2 = temp;
}

int main() {

int num1=50,num2=70;
interchange(&num1,&num2);

printf("\nNumber 1 : %d",num1);
printf("\nNumber 2 : %d",num2);

return(0);
}
```

Output:

```
Number 1 : 70
Number 2 : 50
```

Explanation : Call by Address

- While passing parameter using call by address scheme , we are passing the actual address of the variable to the called function.
- Any updates made inside the called function will modify the original copy since we are directly modifying the content of the exact memory location.

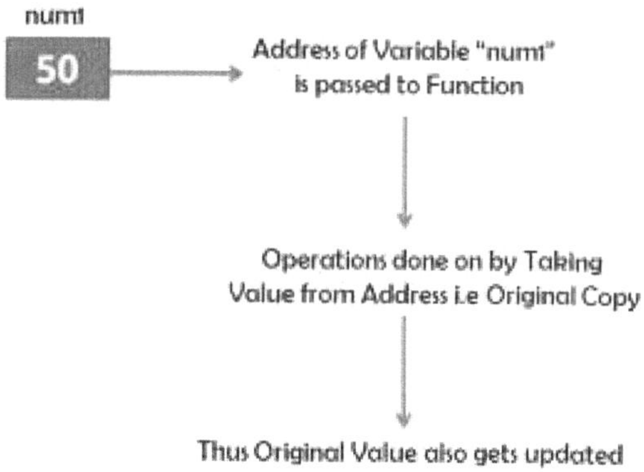

Figure: Pass by reference / pointer / address

Summary of Call By Value and Call By Reference:

Point	**Call by Value**	**Call by Reference**
Copy	Duplicate Copy of Original Parameter is Passed	Actual Copy of Original Parameter is Passed
Modification	No effect on Original Parameter after modifying parameter in function	Original Parameter gets affected if value of parameter changed inside function

3.14. Example Program: Swapping of Two Numbers And Changing The Value Of A Variable Using Pass By Reference

C Program to swap two numbers using pass by reference

```
#include<stdio.h>
#include<conio.h>

void swap(int *num1, int *num2);

void main() {
 int x, y;

 printf("\nEnter First number : ");
 scanf("%d", &x);
```

```
printf("\nEnter Second number : ");
scanf("%d", &y);

printf("\nBefore Swaping x = %d and y = %d", x, y);
swap(&x, &y); // Function Call - Pass By Reference

printf("\nAfter Swaping x = %d and y = %d", x, y);
getch();
}
void swap(int *num1, int *num2) {
 int temp;
 temp = *num1;
 *num1 = *num2;
 *num2 = temp;
}
```

Output

```
Enter First number : 12
Enter Second number : 21

Before Swaping x = 12 and y = 21
After Swaping x = 21 and y = 12
```

Explanation: swap two numbers using pass by reference

In case of pass by reference, we are passing the address of the variables to the function.

swap(&x, &y); // Function Call - Pass By Reference

Consider the above line of code, in this we have passed the address of the variables to the function swap(). In this case while writing the function definition we need to accept those parameters as function argument.

*void swap(int *num1, int *num2)*

See how variable from main function is related with variables from swap function using following table.

Variable in main()	Variables in swap()
Address of Variable 'x'	Equals to num1
Address of Variable 'y'	Equals to num2

UNIT IV

STRUCTURES

Structure - Nested structures - Pointer and Structures - Array of structures - Example Program using structures and pointers - Self referential structures - Dynamic memory allocation - Singly linked list - typedef

4. INTRODUCTION-STRUCTURES

- Arrays allow to define type of variables that can hold several data items of the same kind. Similarly structure is another user defined data type available in C that allows to combine data items of different kinds.
- Structures are used to represent a record. Suppose you want to keep track of your books in a library. You might want to track the following attributes about each book.
  - Title
  - Author
  - Subject
  - Book ID

Definition

Structure is a collection of variables of different types under a single name.

For example

- You want to store some information about a person: his/her name, citizenship number and salary. You can easily create different variables *name, citNo, salary* to store these information separately.
- However, in the future, you would want to store information about multiple persons. Now, you'd need to create different variables for each information per person: *name1, citNo1, salary1, name2, citNo2, salary2*
- You can easily visualize how big and messy the code would look. Also, since no relation between the variables (information) would exist, it's going to be a daunting task.

- A better approach will be to have a collection of all related information under a single name *Person*, and use it for every person. Now, the code looks much cleaner, readable and efficient as well.
- This collection of all related information under a single name Person is a structure.

Structure Definition in C

Keyword ***struct*** is used for creating a structure.

Syntax of structure

```
struct structure_name
{
data_type member1;
data_type member2;
.
.
data_type memeber;
};
```

Note: Don't forget the semicolon}; in the ending line.

We can create the structure for a person as mentioned above as:

```
struct person
{
char name[50];
int citNo;
float salary;
};
```

This declaration above creates the derived data type *struct* **person**.

Structure variable declaration

When a structure is defined, it creates a user-defined type but, no storage or memory is allocated.

For the above structure of a person, variable can be declared as:

```
struct person
{
char name[50];
```

```
int citNo;
float salary;
};

int main()
{
struct person person1, person2, person3[20];
return 0;
}
```

Another way of creating a structure variable is:

```
struct person
{
char name[50];
int citNo;
float salary;
} person1, person2, person3[20];
```

In both cases, two variables *person1*, *person2* and an array *person3* having 20 elements of type **struct person** are created.

Accessing members of a structure

There are two types of operators used for accessing members of a structure.

1. Member operator (.)
2. Structure pointer operator (->)

Any member of a structure can be accessed as:

```
structure_variable_name.member_name
```

Suppose, we want to access salary for variable *person2*. Then, it can be accessed as:

```
person2.salary
```

Example of structure

Write a C program to add two distances entered by user. Measurement of distance should be in inch and feet. (Note: 12 inches = 1 foot)

```
#include <stdio.h>
struct Distance
{
```

```
int feet;
float inch;
} dist1, dist2, sum;
int main()
{
printf("1st distance\n");
// Input of feet for structure variable dist1
printf("Enter feet: ");
scanf("%d", &dist1.feet);
// Input of inch for structure variable dist1
printf("Enter inch: ");
scanf("%f", &dist1.inch);
printf("2nd distance\n");
// Input of feet for structure variable dist2
printf("Enter feet: ");
scanf("%d", &dist2.feet);
// Input of feet for structure variable dist2
printf("Enter inch: ");
scanf("%f", &dist2.inch); sum.feet
= dist1.feet + dist2.feet; sum.inch =
dist1.inch + dist2.inch; if
(sum.inch > 12)
{
        //If inch is greater than 12, changing it to feet.
++sum.feet;
sum.inch = sum.inch - 12;
}
// printing sum of distance dist1 and dist2
printf("Sum of distances = %d\'-%.1f\"", sum.feet, sum.inch);
return 0;
}
```

Output

```
1st distance
Enter feet: 20
Enter inch: 20
2nd distance
Enter feet: 20
Enter inch: 20
Sum of distances = 41'-28.0"
```

4.1. Nested Structure

Structure within Structure: Nested Structure

- Structure written inside another structure is called as nesting of two structures.
- Nested Structures are allowed in C Programming Language.
- We can write one Structure inside another structure as **member** of another structure.

Way 1: Declare two sepa rate structures

```
struct date
{
 int date;
 int month;
 int year;
};

struct Employee
 {
 char ename[20];
 int ssn;
 float salary;
 struct date doj;
}emp1;
```

Accessing Nested Elements:

- Structure members are accessed using **dot operator**.
- **_date_** structure is nested within Employee Structure.
- Members of the **_date_** can be accessed using 'employee'
- **emp1 & doj are two structure names (Variables)**

Explanation of Nested Structure:

```
Accessing Month Field:emp1.doj.month
Accessing day Field:emp1.doj.day
Accessing year Field :emp1.doj.year
```

Way 2: Declare Embedded Structures

```
struct Employee
{
 char ename[20];
 int ssn;
 float salary;
 struct date
 {
 int date;
 int month;
 int year;
 }doj;
}emp1;
```

Accessing Nested Members

Accessing Month Field :emp1.doj.month
Accessing day Field :emp1.doj.day
Accessing year Field:emp1.doj.year

Example

```
#include <stdio.h>
struct Employee
{
 char ename[20];
 int ssn;
 float salary;
 struct date
 {
 int date;
 int month;
 int year;
 }doj;
}emp = {"Pritesh",1000,1000.50,{22,6,1990}};

int main(int argc, char *argv[])
{
printf("\nEmployee Name : %s",emp.ename);
printf("\nEmployee SSN: %d",emp.ssn);
```

```
printf("\nEmployee Salary : %f",emp.salary);
printf("\nEmployee DOJ: %d/%d/%d", \
emp.doj.date,emp.doj.month,emp.doj.year);

return 0;
}
```

Output

```
Employee Name : Pritesh
Employee SSN: 1000
Employee Salary : 1000.500000
Employee DOJ: 22/6/1990
```

4.2. Pointer and Structre

Structures can be created and accessed using pointers. A pointer variable of a structure can be created as below:

```
struct name {
member1;
member2;
.
.
};

int main()
{
struct name *ptr;
}
```

Here, the pointer variable of type ***struct name*** is created.

Accessing structure's member through pointer

A structure's member can be accessed through pointer in two ways:

- Referencing pointer to another address to access memory
- Using dynamic memory allocation

Referencing pointer to another address to access the memory

Consider an example to access structure's member through pointer.

```
#include <stdio.h>
typedef struct person
{
 int age;
```

```
float weight;
};
int main()
{
struct person *personPtr, person1;
personPtr = &person1;// Referencing pointer to memory address of person1
printf("Enter integer: ");
scanf("%d",&(*personPtr).age);
printf("Enter number: ");
scanf("%f",&(*personPtr).weight);
printf("Displaying: ");
printf("%d%f",(*personPtr).age,(*personPtr).weight);
return 0;
}
```

In this example, the pointer variable of type **struct person** is referenced to the address of *person1*. Then, only the structure member through pointer can can accessed.

Using -> operator to access structure pointer member

Structure pointer member can also be accessed using -> operator.

(*personPtr).age is same as personPtr->age

(*personPtr).weight is same as personPtr->weight

Accessing structure member through pointer using dynamic memory allocation

To access structure member using pointers, memory can be allocated dynamically using malloc() function defined under "*stdlib.h*" header file.

Syntax to use malloc()

ptr = (cast-type) malloc(byte-size)*

Example to use structure's member through pointer using malloc() function.

```
#include <stdio.h>
#include <stdlib.h>
struct person {
```

```
 int age;
 float weight;
 char name[30];
};

int main()
{
 struct person *ptr;
 int i, num;

 printf("Enter number of persons: ");
 scanf("%d", &num);

 ptr = (struct person*) malloc(num * sizeof(struct person));
 // Above statement allocates the memory for n structures with pointer
personPtr pointing to base address */

 for(i = 0; i < num; ++i)
 {
 printf("Enter name, age and weight of the person respectively:\n");
 scanf("%s%d%f", &(ptr+i)->name, &(ptr+i)->age, &(ptr+i)->weight);
 }

 printf("Displaying Infromation:\n");
 for(i = 0; i < num; ++i)
 printf("%s\t%d\t%.2f\n", (ptr+i)->name, (ptr+i)->age, (ptr+i)-
>weight);

 return 0;
}
```

Output

```
Enter number of persons: 2
Enter name, age and weight of the person respectively:
Adam
2
3.2
Enter name, age and weight of the person respectively:
```

```
Eve
6
2.3
Displaying Information:
Adam    2  3.20
Eve     6  2.30
```

Example: Calculate Difference between Two Time Periods

```
#include <stdio.h>
struct TIME
{
int  seconds;
int  minutes;
int hours;
};
void differenceBetweenTimePeriod(struct TIME t1, struct TIME t2, struct
TIME *diff);

int main()
{
struct TIME startTime, stopTime, diff;

printf("Enter start time: \n");
printf("Enter hours, minutes and seconds respectively: ");
scanf("%d    %d    %d",    &startTime.hours,    &startTime.minutes,
&startTime.seconds);

printf("Enter stop time: \n");
printf("Enter hours, minutes and seconds respectively: ");
scanf("%d    %d    %d",    &stopTime.hours,    &stopTime.minutes,
&stopTime.seconds);

// Calculate the difference between the start and stop time period.
differenceBetweenTimePeriod(startTime, stopTime, &diff);

printf("\nTIME   DIFFERENCE:  %d:%d:%d  -  ",  startTime.hours,
startTime.minutes, startTime.seconds);
```

```
printf("%d:%d:%d        ",      stopTime.hours,       stopTime.minutes,
stopTime.seconds);
printf("= %d:%d:%d\n", diff.hours, diff.minutes, diff.seconds);

return 0;
}

void differenceBetweenTimePeriod(struct TIME start, struct TIME stop,
struct TIME *diff)
{
if(stop.seconds > start.seconds){
--start.minutes;
start.seconds += 60;
}

diff->seconds = start.seconds - stop.seconds;
if(stop.minutes > start.minutes){
--start.hours;
start.minutes += 60;
}

diff->minutes = start.minutes - stop.minutes;
diff->hours = start.hours - stop.hours;
}
```

Output

```
Enter start time:
Enter hours, minutes and seconds respectively: 12
34
55
Enter stop time:
Enter hours, minutes and seconds respectively:8
12
15

TIME DIFFERENCE: 12:34:55 - 8:12:15 = 4:22:40
```

In this program, user is asked to enter two time periods and these two periods are stored in structure variables *startTime* and *stopTime* respectively.

Then, the function **differenceBetweenTimePeriod** calculates the difference between the time periods and the result is displayed in main() function without returning it (Using call by reference technique).

4.3. Array of Structures

- C Structure is collection of different datatypes (variables) which are grouped together.
- Whereas, array of structures is nothing but collection of structures.
- This is also called as structure array in C.

4.4. Example Program for Array of Structures in C

This program is used to store and access ―id, name and percentage‖ for 3 students. Structure array is used in this program to store and display records for many students. You can store ―n‖ number of students record by declaring structure variable as ‗struct student record[n]―, where n can be 1000 or 5000 etc.

```
#include <stdio.h>
#include <string.h>

struct student
{
 int id;
 char name[30];
 float percentage;
};

int main()
{
 int i;
 struct student record[2];

 // 1st student's record
 record[0].id=1;
```

```
strcpy(record[0].name, "Raju");
record[0].percentage = 86.5;

// 2nd student's record
record[1].id=2;
strcpy(record[1].name, "Surendren");
record[1].percentage = 90.5;

// 3rd student's record
record[2].id=3;
strcpy(record[2].name, "Thiyagu");
record[2].percentage = 81.5;

for(i=0; i<3; i++)
{
printf(" Records of STUDENT : %d \n", i+1);
printf(" Id is: %d \n", record[i].id);
printf(" Name is: %s \n", record[i].name);
printf(" Percentage is: %f\n\n",record[i].percentage);
}
return 0;
}
```

Output

```
Records of STUDENT : 1
Id is: 1
Name is: Raju
Percentage is: 86.500000
Records of STUDENT : 2
Id is: 2
Name is: Surendren
Percentage is: 90.500000
Records of STUDENT : 3
Id is: 3
Name is: Thiyagu
Percentage is: 81.500000
```

4.5. Example Program for Declaring Many Structure Variable in C

In this program, two structure variables —record1" and –record2" are declared for same structure and different values are assigned for both structure variables. Separate memory is allocated for both structure variables to store the data.

```
#include <stdio.h>
#include <string.h>

struct student
{
 int id;
 char name[30];
 float percentage;
};

int main()
{
 int i;
 struct student record1 = {1, "Raju", 90.5};
 struct student record2 = {2, "Mani", 93.5};

 printf("Records of STUDENT1: \n");
 printf("Id is: %d \n", record1.id);
 printf("Name is: %s \n", record1.name);
 printf("Percentage is: %f \n\n", record1.percentage);

 printf("Records of STUDENT2: \n");
 printf("Id is: %d \n", record2.id);
 printf("Name is: %s \n", record2.name);
 printf("Percentage is: %f \n\n", record2.percentage);

 return 0;
}
```

Output

```
Records of STUDENT1:
Id is: 1
Name is: Raju
Percentage is: 90.500000
Records of STUDENT2:
```

```
Id is: 2
Name is: Mani
Percentage is: 93.500000
```

4.6. Self Referential Structures

- Let us understand what is a self-referential structure in C programming and how do we implement a self referential structure in data structures.
- Before we go ahead with self-referential data structures, let us also have a look at what are structures and in case you know it, you can skip it.

What are Structures?

- An array is a collection of homogeneous elements but in the real world, we may need to include different types of logically related data.
- To store different data such as name, age, contact, etc. associated with a single entity, we make use of structures.
- A structure is, therefore, capable of storing heterogeneous data under a single name and the data elements are called as members.

What is a Self-Referential Structure?

- A structure that contains pointers to a structure of its own type is known as self-referential structure.
- In other words, a self-referential C structure is the one which includes a pointer to an instance of itself.

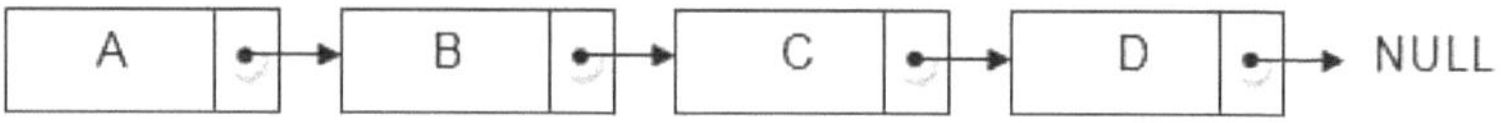

Figure: Self Referential Structures

```
struct demo
{
datatype member1, member2;
struct demo *ptr1, *ptr2;
}
```

As you can see in the syntax, ptr1 and ptr2 are structure pointers that are pointing to the structure demo, so structure demo is a self referential structure. These types of data structures are helpful in implementing data structures like linked lists and trees.

It is an error to use a structure variable as a member of its own struct type structure or union type union, respectively.

Self Referential Structure Example

```
struct node
{
int data;
struct node *link;
};
```

- The concept of linked lists, stacks, queues, trees and many others works on the principle of self-referential structures.
- One important point worth noting is that you cannot reference the **typedef** that you create within the structure itself in C programming.

Implementation of Self Referential Structure in Linked Lists

```
#include<stdio.h>
#include<stdlib.h>

struct node
{
int info;
struct node *link;
}*front = NULL, *rear = NULL;

int isEmpty()
{
if(front == NULL)
{
return 1;
}
else
{
```

```
return 0;
}
}

void insert(int element)
{
struct node *t;
t = (struct node *)malloc(sizeof(struct node));
if(t == NULL)
{
printf("Queue is not Allocated\n");
return;
}
t->info = element;
t->link = NULL;
if(front == NULL)
{
front = t;
}
else
{
rear->link = t;
}
rear = t;
}

void display()
{
struct node *ptr;
if(isEmpty())
{
printf("Empty\n");
return;
}
printf("Queue Elements:\n");
for(ptr = front; ptr != NULL; ptr = ptr->link)
{
```

```
printf("%d ", ptr->info);
}
printf("\n\n");
}

int main()
{
int option, element;
while(1)
{
printf("1. Insert an Element (Node) in the Queue\n");
printf("2. Display All Elements of the queue\n");
printf("3. Exit\n");
printf("Enter your option:\t");
scanf("%d", &option);
switch(option)
{
case 1:
printf("Enter the Element to Add in Queue:\t");
scanf("%d", &element);
insert(element);
break;
case 2:
display();
break;
case 3:
exit(1);
default:
printf("Please select an appropriate option\n");
}
}
return 0;
}
```

4.7. Dynamic Memory Allocation

- You have to declare the size of an array before you use it. Hence, the array you declared may be insufficient or more than required to hold data. To solve this issue, you can allocate memory dynamically.

- Dynamic memory management refers to manual memory management. This allows you to obtain more memory when required and release it when not necessary.
- Although C inherently does not have any technique to allocate memory dynamically, there are 4library functions defined under <*stdlib.h*> for dynamic memory allocation.

Function	Use of Function
malloc()	Allocates requested size of bytes and returns a pointer first byte of allocated space
calloc()	Allocates space for an array elements, initializes to zero and then returns a pointer to memory
free()	deallocate the previously allocated space
realloc()	Change the size of previously allocated space

C malloc()

The name malloc stands for "memory allocation".

The function malloc() reserves a block of memory of specified size and return a pointer of type void which can be casted into pointer of any form.

Syntax of malloc()

ptr = (cast-type) malloc(byte-size)*

Here, *ptr* is pointer of cast-type. The malloc() function returns a pointer to an area of memory with size of byte size. If the space is insufficient, allocation fails and returns NULL pointer.

ptr = (int) malloc(100 * sizeof(int));*

This statement will allocate either 200 or 400 according to size of int 2 or 4 bytes respectively and the pointer points to the address of first byte of memory.

C calloc()

The name calloc stands for "contiguous allocation".

The only difference between malloc() and calloc() is that, malloc() allocates single block of memory whereas calloc() allocates multiple blocks of memory each of same size and sets all bytes to zero.

Syntax of calloc()

ptr = (cast-type)calloc(n, element-size);*

This statement will allocate contiguous space in memory for an array of *n* elements. For example:

ptr = (float) calloc(25, sizeof(float));*

This statement allocates contiguous space in memory for an array of 25 elements each of size of float, i.e, 4 bytes.

C free()

Dynamically allocated memory created with either calloc() or malloc() doesn't get freed on its own. You must explicitly use free() to release the space.

syntax of free()

free(ptr);

This statement frees the space allocated in the memory pointed by ptr.

Example #1: Using C malloc() and free()

Write a C program to find sum of n elements entered by user. To perform this program, allocate memory dynamically using malloc() function.

```
#include <stdio.h>
#include <stdlib.h>
int main()
{
int num, i, *ptr, sum = 0;
printf("Enter number of elements: ");
scanf("%d", &num);
ptr = (int*) malloc(num * sizeof(int));//memory allocated using malloc
if(ptr == NULL)
{
printf("Error! memory not allocated.");
exit(0);
}
printf("Enter elements of array: ");
for(i = 0; i < num; ++i)
{
```

```
scanf("%d", ptr + i);
sum += *(ptr + i);
}
printf("Sum = %d", sum);
free(ptr);
return 0;
}
```

Output:

```
Enter number of elements: 4
Enter elements of array: 2 5 6 7
Sum = 20
```

Example #2: Using C calloc() and free()

Write a C program to find sum of n elements entered by user. To perform this program, allocate memory dynamically using calloc() function.

```
#include <stdio.h>
#include <stdlib.h>
int main()
{
int num, i, *ptr, sum = 0;
printf("Enter number of elements: ");
scanf("%d", &num);
ptr = (int*) calloc(num, sizeof(int));
if(ptr == NULL)
{
printf("Error! memory not allocated.");
exit(0);
}
printf("Enter elements of array: ");
for(i = 0; i < num; ++i)
{
scanf("%d", ptr + i);
sum += *(ptr + i);
}
printf("Sum = %d", sum);
free(ptr);
return 0;
}
```

Output

Enter number of elements: 4
Enter elements of array: 5 6 2 4
Sum = 17

C realloc()

If the previously allocated memory is insufficient or more than required, you can change the previously allocated memory size using realloc().

Syntax of realloc()

ptr = realloc(ptr, newsize);

Here, *ptr* is reallocated with size of newsize.

Example #3: Using realloc()

```
#include <stdio.h>
#include <stdlib.h>
int main()
{
int *ptr, i , n1, n2;
printf("Enter size of array: ");
scanf("%d", &n1);

ptr = (int*) malloc(n1 * sizeof(int));

printf("Address of previously allocated memory: ");
for(i = 0; i < n1; ++i)
printf("%u\t",ptr + i);
printf("\nEnter new size of array: ");
scanf("%d", &n2);
ptr = realloc(ptr, n2);
for(i = 0; i < n2; ++i)
printf("%u\t", ptr + i);
return 0;
}
```

Output

Enter size of array: 3

Address of previously allocated memory: 9199632 9199636 9199640
Enter new size of array: 5
9199632 9199636 9199640 9199644 9199648

4.8. Singly Linked List

A linked list is a way to store a collection of elements. Like an array these can be character or integers. Each element in a linked list is stored in the form of a node.

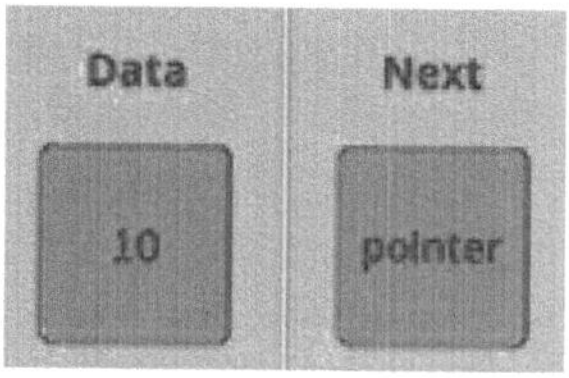

Figure: Node

A node is a collection of two sub-elements or parts. A data part that stores the element and a next part that stores the link to the next node.

Linked List:

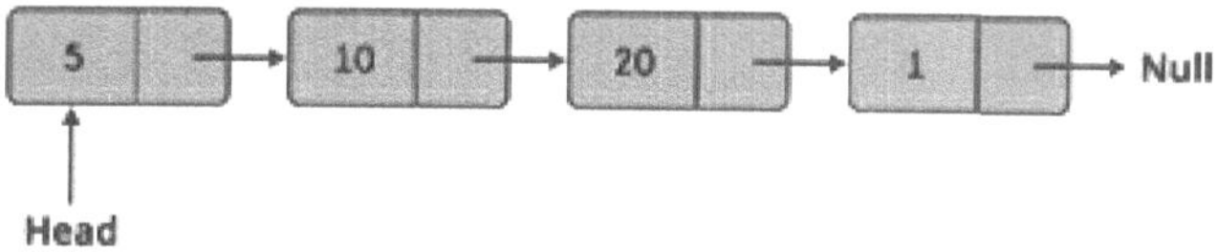

Figure: Linked List

A linked list is formed when many such nodes are linked together to form a chain. Each node points to the next node present in the order. The first node is always used as a reference to traverse the list and is called HEAD. The last node points to NULL.

Declaring a Linked list

In C language, a linked list can be implemented using structure and pointers.

```
struct LinkedList{
int data;
struct LinkedList *next;
 };
```

The above definition is used to create every node in the list. The **data** field stores the element and the **next** is a pointer to store the address of the next node.

Noticed something unusual with next?

In place of a data type, **struct LinkedList** is written before next. That's because its a **self-referencing pointer**. It means a pointer that points to whatever it is a part of. Here next is a part of a node and it will point to the next node.

Creating a Node

Let's define a data type of struct LinkedListto make code cleaner.

```
typedef struct LinkedList *node; //Define node as pointer of data type struct
LinkedList

node createNode(){
node temp; // declare a node
temp = (node)malloc(sizeof(struct LinkedList)); // allocate memory using
malloc()
temp->next = NULL;// make next point to NULL
return temp;//return the new node
}
```

- **typedef** is used to define a data type in C.
- **malloc()** is used to dynamically allocate a single block of memory in C, it is available in the header file stdlib.h.
- **sizeof()** is used to determine size in bytes of an element in C. Here it is used to determine size of each node and sent as a parameter to malloc.

The above code will create a node with data as value and next pointing to NULL.

Let's see how to **add a node to the linked list**:

```
node addNode(node head, int value){
node temp,p;// declare two nodes temp and p
temp = createNode();//createNode will return a new node with data = value
and next pointing to NULL.
temp->data = value; // add element's value to data part of node
```

```
if(head == NULL){
head = temp; //when linked list is empty
}
else{
p= head;//assign head to p
while(p->next != NULL){
p = p->next;//traverse the list until p is the last node.The last node always points to NULL.
}
p->next = temp;//Point the previous last node to the new node created.
}
return head;
}
```

Here the new node will always be added after the last node. This is known as inserting a node at the rear end.

Food for thought

This type of linked list is known as simple or singly linked list. A simple linked list can be traversed in only one direction from head to the last node.

The last node is checked by the condition

```
p->next = NULL;
```

Here -> is used to access **next** sub element of node p. **NULL** denotes no node exists after the current node , i.e. its the end of the list.

Traversing the list:

The linked list can be traversed in a while loop by using the head node as a starting reference:

```
node p;
p = head;
while(p != NULL){
p = p->next;
}
```

C Program to Implement Singly Linked List using Dynamic Memory Allocation

- This C Program Implements Singly Linked List using Dynamic Memory Allocation.
- Here is source code of the C Program to Implement Singly Linked List using Dynamic Memory Allocation. The C program is successfully compiled and run on a Linux system. The program output is also shown below.

```
/*
 * C Program to Implement Singly Linked List using Dynamic Memory
Allocation
 */
#include <stdio.h>
#include <malloc.h>
#define ISEMPTY printf("\nEMPTY LIST:");
/*
 * Node Declaration
 */
struct node
{
int value;
struct node *next;
};

snode* create_node(int);
void insert_node_first();
void insert_node_last();
void insert_node_pos();
void sorted_ascend();
void delete_pos();
void search();
void update_val();
void display();
void rev_display(snode *);

typedef struct node snode;
snode *newnode, *ptr, *prev, *temp;
```

```
snode *first = NULL, *last = NULL;

/*
 * Main :contains menu
 */

 int main()
 {
int ch;
char ans = 'Y';

while (ans == 'Y' | |ans == 'y')
{
printf("\n--------------------------------\n");
printf("\nOperations on singly linked list\n");
printf("\n--------------------------------\n");
printf("\n1.Insert node at first");
printf("\n2.Insert node at last");
printf("\n3.Insert node at position");
printf("\n4.Sorted Linked List in Ascending Order");
printf("\n5.Delete Node from any Position");
printf("\n6.Update Node Value");
printf("\n7.Search Element in the linked list"); printf("\n8.Display
List from Beginning to end"); printf("\n9.Display List from end
using Recursion"); printf("\n10.Exit\n");
printf("\n~~~~~~~~~~~~~~~~~~~~~~~~~~~~~~~~~~~~~~~~~\n");
printf("\nEnter your choice");
scanf("%d", &ch);

switch (ch)
{
case 1:
printf("\n...Inserting node at first...\n");
insert_node_first();
break;
case 2:
```

```
printf("\n...Inserting node at last...\n");
insert_node_last();
break;
case 3:
printf("\n...Inserting node at position...\n");
insert_node_pos();
break;
case 4:
printf("\n...Sorted Linked List in Ascending Order...\n");
sorted_ascend();
break;
case 5:
printf("\n...Deleting Node from any Position...\n");
delete_pos();
break;
case 6:
printf("\n...Updating Node Value...\n");
update_val();
break;
case 7:
printf("\n...Searching Element in the List...\n");
search();
break;
case 8:
printf("\n...Displaying List From Beginning to End...\n");
display();
break;
case 9:
printf("\n...Displaying List From End using Recursion...\n");
rev_display(first);
break; case 10:
printf("\n...Exiting...\n");
return 0;
break;
default:
printf("\n...Invalid Choice...\n");
```

```
break;
}
printf("\nYOU WANT TO CONTINUE (Y/N)");
scanf(" %c", &ans);
}
return 0;
 }

/*
 * Creating Node
 */
snode* create_node(int val)
{
newnode = (snode *)malloc(sizeof(snode));
if (newnode == NULL)
{
printf("\nMemory was not allocated");
return 0;
}
else
{
newnode->value = val;
newnode->next = NULL;
return newnode;
}
}

/*
 * Inserting Node at First
 */
void insert_node_first()
{
int val;

printf("\nEnter the value for the node:");
scanf("%d", &val);
newnode = create_node(val);
```

```
if (first == last && first == NULL)
{
first = last = newnode;
first->next = NULL;
last->next = NULL;
}
else
{
temp = first;
first = newnode;
first->next = temp;
}
printf("\n----INSERTED----");
}

/*
 * Inserting Node at Last
 */
void insert_node_last()
{
int val;

printf("\nEnter the value for the Node:");
scanf("%d", &val);
newnode = create_node(val);
if (first == last && last == NULL)
{
first = last = newnode;
first->next = NULL;
last->next = NULL;
}
else
{
last->next = newnode;
last = newnode;
last->next = NULL;
}
```

```
printf("\n----INSERTED----");
}

/*
 * Inserting Node at position
 */
void insert_node_pos()
{
int pos, val, cnt = 0, i;

printf("\nEnter the value for the Node:");
scanf("%d", &val);
newnode = create_node(val);
printf("\nEnter the position ");
scanf("%d", &pos);
ptr = first;
while (ptr != NULL)
{
ptr = ptr->next;
cnt++;
}
if (pos == 1)
{
if (first == last && first == NULL)
{
first = last = newnode;
first->next = NULL;
last->next = NULL;
}
else
{
temp = first;
first = newnode;
first->next = temp;
}
printf("\nInserted");
}
```

```
else if (pos>1 && pos<=cnt)
{
ptr = first;
for (i = 1;i < pos;i++)
{
prev = ptr;
ptr = ptr->next;
}
prev->next = newnode;
newnode->next = ptr;
printf("\n----INSERTED----");
}
else
{
printf("Position is out of range");
}
}

/*
 * Sorted Linked List
 */
void sorted_ascend()
{
snode *nxt;
int t;

if (first == NULL)
{
ISEMPTY;
printf(":No elements to sort\n");
}
else
{
for (ptr = first;ptr != NULL;ptr = ptr->next)
{
for (nxt = ptr->next;nxt != NULL;nxt = nxt->next)
{
```

```
if (ptr->value > nxt->value)
{
t = ptr->value;
ptr->value = nxt->value;
nxt->value = t;
}
}
}
printf("\n---Sorted List---");
for (ptr = first;ptr != NULL;ptr = ptr->next)
{
printf("%d\t", ptr->value);
}
}
}

/*
 * Delete Node from specified position in a non-empty list
 */
void delete_pos()
{
int pos, cnt = 0, i;

if (first == NULL)
{
ISEMPTY;
printf(":No node to delete\n");
}
else
{
printf("\nEnter the position of value to be deleted:");
scanf(" %d", &pos);
ptr = first;
if (pos == 1)
{
first = ptr->next;
printf("\nElement deleted");
```

```
}
else
{
while (ptr != NULL)
{
ptr = ptr->next;
cnt = cnt + 1;
}
if (pos > 0 && pos <= cnt)
{
ptr = first;
for (i = 1;i < pos;i++)
{
prev = ptr;
ptr = ptr->next;
}
prev->next = ptr->next;
}
else
{
printf("Position is out of range");
}
free(ptr);
printf("\nElement deleted");
}
}
}
/*
 * Updating Node value in a non-empty list
 */
void update_val()
{
int oldval, newval, flag = 0;

if (first == NULL)
{
ISEMPTY;
```

```
printf(":No nodes in the list to update\n");
}
else
{
printf("\nEnter the value to be updated:");
scanf("%d", &oldval);
printf("\nEnter the newvalue:");
scanf("%d", &newval);
for (ptr = first;ptr != NULL;ptr = ptr->next)
{
if (ptr->value == oldval)
{
ptr->value = newval;
flag = 1;
break;
}
}
if (flag == 1)
{
printf("\nUpdated Successfully");
}
else
{
printf("\nValue not found in List");
}
}
}

/*
 * searching an element in a non-empty list
 */
void search()
{
int flag = 0, key, pos = 0;

if (first == NULL)
{
```

```
ISEMPTY;
printf(":No nodes in the list\n");
}
else
{
printf("\nEnter the value to search");
scanf("%d", &key);
for (ptr = first;ptr != NULL;ptr = ptr->next)
{
pos = pos + 1;
if (ptr->value == key)
{
flag = 1;
break;
}
}
if (flag == 1)
{
printf("\nElement %d found at %d position\n", key, pos);
}
else
{
printf("\nElement %d not found in list\n", key);
}
}
}
/*
 * Displays non-empty List from Beginning to End
 */
void display()
{
if (first == NULL)
{
ISEMPTY;
printf(":No nodes in the list to display\n");
}
else
```

```
{
for (ptr = first;ptr != NULL;ptr = ptr->next)
{
printf("%d\t", ptr->value);
}
}
}

/*
 * Display non-empty list in Reverse Order
 */
void rev_display(snode *ptr)
{
int val;

if (ptr == NULL)
{
ISEMPTY;
printf(":No nodes to display\n");
}
else
{
if (ptr != NULL)
{
val = ptr->value;
rev_display(ptr->next);
printf("%d\t", val);
}

}
}
```

Output

```
$gcc linkedlist.c
a.out
-------------------------------
 Operations on singly linked list
 -------------------------------
```

```
 1.Insert node at first
2.Insert node at last
3.Insert node at position
4.Sorted Linked List in Ascending Order
5.Delete Node from any Position
6.Update Node Value
7.Search Element in the linked list
8.Display List from Beginning to end
9.Display List from end using Recursion
10.Exit
 ~~~~~~~~~~~~~~~~~~~~~~~~~~~~~~~~~~~~~
 Enter your choice8

...Displaying List From Beginning to End...

EMPTY LIST::No nodes in the list to display

YOU WANT TO CONTINUE (Y/N)y
 --------------------------------
 Operations on singly linked list
 --------------------------------
 1.Insert node at first
2.Insert node at last
3.Insert node at position
4.Sorted Linked List in Ascending Order
5.Delete Node from any Position
6.Update Node Value
7.Search Element in the linked list
8.Display List from Beginning to end
9.Display List from end using Recursion
10.Exit
 ~~~~~~~~~~~~~~~~~~~~~~~~~~~~~~~~~~~~~
 Enter your choice5

...Deleting Node from any Position...

EMPTY LIST::No node to delete
```

```
YOU WANT TO CONTINUE (Y/N)
y
 --------------------------------
 Operations on singly linked list
 --------------------------------
 1.Insert node at first
2.Insert node at last
3.Insert node at position
4.Sorted Linked List in Ascending Order
5.Delete Node from any Position
6.Update Node Value
7.Search Element in the linked list
8.Display List from Beginning to end
9.Display List from end using Recursion
10.Exit
 ~~~~~~~~~~~~~~~~~~~~~~~~~~~~~~~~~~~~~~
 Enter your choice6

...Updating Node Value...

EMPTY LIST::No nodes in the list to update

YOU WANT TO CONTINUE (Y/N)y
 --------------------------------
 Operations on singly linked list


--------------------------------
 1.Insert node at first
2.Insert node at last
3.Insert node at position
4.Sorted Linked List in Ascending Order
5.Delete Node from any Position
6.Update Node Value
7.Search Element in the linked list
8.Display List from Beginning to end
9.Display List from end using Recursion
```

```
10.Exit
~~~~~~~~~~~~~~~~~~~~~~~~~~~~~~~~~~~
Enter your choice
7

...Searching Element in the List...

EMPTY LIST::No nodes in the list

YOU WANT TO CONTINUE (Y/N)y
--------------------------------
Operations on singly linked list
--------------------------------
1.Insert node at first
2.Insert node at last
3.Insert node at position
4.Sorted Linked List in Ascending Order
5.Delete Node from any Position
6.Update Node Value
7.Search Element in the linked list
8.Display List from Beginning to end
9.Display List from end using Recursion
10.Exit
~~~~~~~~~~~~~~~~~~~~~~~~~~~~~~~~~~~
Enter your choice
3

...Inserting node at position...

Enter the value for the Node:1010

Enter the position 5
Position is out of range
YOU WANT TO CONTINUE (Y/N)
--------------------------------
Operations on singly linked list
--------------------------------
```

```
 1.Insert node at first
2.Insert node at last
3.Insert node at position
4.Sorted Linked List in Ascending Order
5.Delete Node from any Position
6.Update Node Value
7.Search Element in the linked list
8.Display List from Beginning to end
9.Display List from end using Recursion
10.Exit
 ~~~~~~~~~~~~~~~~~~~~~~~~~~~~~~~~~~~~~~~~
 Enter your choice1

...Inserting node at first...

Enter the value for the node:100
 ----INSERTED----
YOU WANT TO CONTINUE (Y/N)y
 --------------------------------
 Operations on singly linked list
 --------------------------------
 1.Insert node at first
2.Insert node at last
3.Insert node at position
4.Sorted Linked List in Ascending Order
5.Delete Node from any Position
6.Update Node Value
7.Search Element in the linked list
8.Display List from Beginning to end
9.Display List from end using Recursion
10.Exit
 ~~~~~~~~~~~~~~~~~~~~~~~~~~~~~~~~~~~~~~~~
 Enter your choice1

...Inserting node at first...

Enter the value for the node:200
```

```
----INSERTED----
YOU WANT TO CONTINUE (Y/N)y
 --------------------------------
 Operations on singly linked list
 --------------------------------
 1.Insert node at first
2.Insert node at last
3.Insert node at position
4.Sorted Linked List in Ascending Order
5.Delete Node from any Position
6.Update Node Value
7.Search Element in the linked list
8.Display List from Beginning to end
9.Display List from end using Recursion
10.Exit
 ~~~~~~~~~~~~~~~~~~~~~~~~~~~~~~~~~~~~~~
 Enter your choice8

...Displaying List From Beginning to End...
200 100
YOU WANT TO CONTINUE (Y/N)y
 --------------------------------


Operations on singly linked list
 --------------------------------
 1.Insert node at first
2.Insert node at last
3.Insert node at position
4.Sorted Linked List in Ascending Order
5.Delete Node from any Position
6.Update Node Value
7.Search Element in the linked list
8.Display List from Beginning to end
9.Display List from end using Recursion
10.Exit
 ~~~~~~~~~~~~~~~~~~~~~~~~~~~~~~~~~~~~~~
```

```
Enter your choice2

...Inserting node at last...

Enter the value for the Node:50

----INSERTED----
YOU WANT TO CONTINUE (Y/N)y
 -------------------------------
 Operations on singly linked list
 -------------------------------
 1.Insert node at first
2.Insert node at last
3.Insert node at position
4.Sorted Linked List in Ascending Order
5.Delete Node from any Position
6.Update Node Value
7.Search Element in the linked list
8.Display List from Beginning to end
9.Display List from end using Recursion
10.Exit

~~~~~~~~~~~~~~~~~~~~~~~~~~~~~~~~~~~~~~~

Enter your choice2

...Inserting node at last...

Enter the value for the Node:150

----INSERTED----
YOU WANT TO CONTINUE (Y/N)y
-------------------------------
Operations on singly linked list
-------------------------------
1.Insert node at first
2.Insert node at last
```

```
3.Insert node at position
4.Sorted Linked List in Ascending Order
5.Delete Node from any Position
6.Update Node Value
7.Search Element in the linked list
8.Display List from Beginning to end
9.Display List from end using Recursion
10.Exit
~~~~~~~~~~~~~~~~~~~~~~~~~~~~~~~~~~~~~~~~
Enter your choice8

...Displaying List From Beginning to End...
200 100 50150
YOU WANT TO CONTINUE (Y/N)
--------------------------------
Operations on singly linked list
--------------------------------
1.Insert node at first
2.Insert node at last
3.Insert node at position
4.Sorted Linked List in Ascending Order
5.Delete Node from any Position
6.Update Node Value
7.Search Element in the linked list
8.Display List from Beginning to end
9.Display List from end using Recursion
10.Exit
~~~~~~~~~~~~~~~~~~~~~~~~~~~~~~~~~~~~~~~~
Enter your choice3

...Inserting node at position...

Enter the value for the Node:1111

Enter the position 4

----INSERTED----
```

```
YOU WANT TO CONTINUE (Y/N)y
--------------------------------
Operations on singly linked list
--------------------------------
1.Insert node at first
2.Insert node at last
3.Insert node at position
4.Sorted Linked List in Ascending Order
5.Delete Node from any Position
6.Update Node Value
7.Search Element in the linked list
8.Display List from Beginning to end
9.Display List from end using Recursion
10.Exit
~~~~~~~~~~~~~~~~~~~~~~~~~~~~~~~~~~~~~~
Enter your choice8

...Displaying List From Beginning to End...
200 100 501111150
YOU WANT TO CONTINUE (Y/N)
y
--------------------------------
Operations on singly linked list
--------------------------------
1.Insert node at first
2.Insert node at last
3.Insert node at position
4.Sorted Linked List in Ascending Order
5.Delete Node from any Position
6.Update Node Value
7.Search Element in the linked list
8.Display List from Beginning to end
9.Display List from end using Recursion
10.Exit
~~~~~~~~~~~~~~~~~~~~~~~~~~~~~~~~~~~~~~
Enter your choice3
```

```
...Inserting node at position...

Enter the value for the Node:1010

Enter the position 100
Position is out of range
YOU WANT TO CONTINUE (Y/N)y
--------------------------------
Operations on singly linked list
--------------------------------
1.Insert node at first
2.Insert node at last
3.Insert node at position
4.Sorted Linked List in Ascending Order
5.Delete Node from any Position
6.Update Node Value
7.Search Element in the linked list
8.Display List from Beginning to end
9.Display List from end using Recursion
10.Exit
~~~~~~~~~~~~~~~~~~~~~~~~~~~~~~~~~~~~~
Enter your choice8

...Displaying List From Beginning to End...
200 100 501111150
YOU WANT TO CONTINUE (Y/N)y
--------------------------------
Operations on singly linked list
--------------------------------
1.Insert node at first
2.Insert node at last
3.Insert node at position
4.Sorted Linked List in Ascending Order
5.Delete Node from any Position
6.Update Node Value
7.Search Element in the linked list
8.Display List from Beginning to end
```

```
9.Display List from end using Recursion
10.Exit
~~~~~~~~~~~~~~~~~~~~~~~~~~~~~~~~~~~~~~~~~
Enter your choice5

...Deleting Node from any Position...

Enter the position of value to be deleted:1

Element deleted
YOU WANT TO CONTINUE (Y/N)y
--------------------------------
Operations on singly linked list
--------------------------------
1.Insert node at first
2.Insert node at last
3.Insert node at position
4.Sorted Linked List in Ascending Order
5.Delete Node from any Position
6.Update Node Value
7.Search Element in the linked list
8.Display List from Beginning to end
9.Display List from end using Recursion
10.Exit
~~~~~~~~~~~~~~~~~~~~~~~~~~~~~~~~~~~~~~~~~
Enter your choice8

...Displaying List From Beginning to End...
100 501111150
YOU WANT TO CONTINUE (Y/N)y
--------------------------------
Operations on singly linked list
--------------------------------
1.Insert node at first
2.Insert node at last
3.Insert node at position
4.Sorted Linked List in Ascending Order
```

```
5.Delete Node from any Position
6.Update Node Value
7.Search Element in the linked list
8.Display List from Beginning to end
9.Display List from end using Recursion
10.Exit
~~~~~~~~~~~~~~~~~~~~~~~~~~~~~~~~~~~~~
Enter your choice5

...Deleting Node from any Position...

Enter the position of value to be deleted:4

Element deleted
YOU WANT TO CONTINUE (Y/N)y
--------------------------------
Operations on singly linked list


--------------------------------
1.Insert node at first
2.Insert node at last
3.Insert node at position
4.Sorted Linked List in Ascending Order
5.Delete Node from any Position
6.Update Node Value
7.Search Element in the linked list
8.Display List from Beginning to end
9.Display List from end using Recursion
10.Exit
~~~~~~~~~~~~~~~~~~~~~~~~~~~~~~~~~~~~~
Enter your choice8

...Displaying List From Beginning to End...
100 501111

YOU WANT TO CONTINUE (Y/N)y
--------------------------------
```

```
Operations on singly linked list
--------------------------------
1.Insert node at first
2.Insert node at last
3.Insert node at position
4.Sorted Linked List in Ascending Order
5.Delete Node from any Position
6.Update Node Value
7.Search Element in the linked list
8.Display List from Beginning to end
9.Display List from end using Recursion
10.Exit
~~~~~~~~~~~~~~~~~~~~~~~~~~~~~~~~~~~~~~
Enter your choice5

...Deleting Node from any Position...

Enter the position of value to be deleted:2

Element deleted
YOU WANT TO CONTINUE (Y/N)y
--------------------------------
Operations on singly linked list
--------------------------------
1.Insert node at first
2.Insert node at last
3.Insert node at position
4.Sorted Linked List in Ascending Order
5.Delete Node from any Position
6.Update Node Value
7.Search Element in the linked list
8.Display List from Beginning to end
9.Display List from end using Recursion
10.Exit
~~~~~~~~~~~~~~~~~~~~~~~~~~~~~~~~~~~~~~
Enter your choice8
```

```
...Displaying List From Beginning to End...
100 1111
YOU WANT TO CONTINUE (Y/N)y
-------------------------------
Operations on singly linked list
-------------------------------
1.Insert node at first
2.Insert node at last
3.Insert node at position
4.Sorted Linked List in Ascending Order
5.Delete Node from any Position
6.Update Node Value
7.Search Element in the linked list
8.Display List from Beginning to end
9.Display List from end using Recursion
10.Exit
~~~~~~~~~~~~~~~~~~~~~~~~~~~~~~~~~~~~
Enter your choice6

...Updating Node Value...

Enter the value to be updated:100

Enter the newvalue:10101

Updated Successfully
YOU WANT TO CONTINUE (Y/N)y
-------------------------------
Operations on singly linked list
-------------------------------
1.Insert node at first
2.Insert node at last
3.Insert node at position
4.Sorted Linked List in Ascending Order
5.Delete Node from any Position
6.Update Node Value
7.Search Element in the linked list
```

```
8.Display List from Beginning to end
9.Display List from end using Recursion
10.Exit
~~~~~~~~~~~~~~~~~~~~~~~~~~~~~~~~~~~~~~~
Enter your choice8

...Displaying List From Beginning to End...
10101 1111
YOU WANT TO CONTINUE (Y/N)y
--------------------------------
Operations on singly linked list
--------------------------------
1.Insert node at first
2.Insert node at last
3.Insert node at position
4.Sorted Linked List in Ascending Order
5.Delete Node from any Position
6.Update Node Value
7.Search Element in the linked list
8.Display List from Beginning to end
9.Display List from end using Recursion
10.Exit
~~~~~~~~~~~~~~~~~~~~~~~~~~~~~~~~~~~~~~~
Enter your choice6

...Updating Node Value...

Enter the value to be updated:100

Enter the newvalue:200

Value not found in List
YOU WANT TO CONTINUE (Y/N)y
--------------------------------
Operations on singly linked list
--------------------------------
1.Insert node at first
```

```
2.Insert node at last
3.Insert node at position
4.Sorted Linked List in Ascending Order
5.Delete Node from any Position
6.Update Node Value
7.Search Element in the linked list
8.Display List from Beginning to end
9.Display List from end using Recursion
10.Exit
~~~~~~~~~~~~~~~~~~~~~~~~~~~~~~~~~~~~~~
Enter your choice7

...Searching Element in the List...

Enter the value to search 1111

Element 1111 found at 2 position

YOU WANT TO CONTINUE (Y/N)y
--------------------------------
Operations on singly linked list
--------------------------------
1.Insert node at first
2.Insert node at last
3.Insert node at position
4.Sorted Linked List in Ascending Order
5.Delete Node from any Position
6.Update Node Value
7.Search Element in the linked list
8.Display List from Beginning to end
9.Display List from end using Recursion
10.Exit
~~~~~~~~~~~~~~~~~~~~~~~~~~~~~~~~~~~~~~
Enter your choice7

...Searching Element in the List...
```

```
Enter the value to search200

Element 200 not found in list

YOU WANT TO CONTINUE (Y/N)y
--------------------------------
Operations on singly linked list
--------------------------------
1.Insert node at first
2.Insert node at last
3.Insert node at position
4.Sorted Linked List in Ascending Order
5.Delete Node from any Position
6.Update Node Value
7.Search Element in the linked list
8.Display List from Beginning to end
9.Display List from end using Recursion
10.Exit
~~~~~~~~~~~~~~~~~~~~~~~~~~~~~~~~~~~~~~
Enter your choice4

...Sorted Linked List in Ascending Order...

---Sorted List---1111 10101
YOU WANT TO CONTINUE (Y/N)y
--------------------------------
Operations on singly linked list
 --------------------------------
 1.Insert node at first
2.Insert node at last
3.Insert node at position
4.Sorted Linked List in Ascending Order
5.Delete Node from any Position
6.Update Node Value
7.Search Element in the linked list
8.Display List from Beginning to end
9.Display List from end using Recursion
```

```
10.Exit
~~~~~~~~~~~~~~~~~~~~~~~~~~~~~~~~~~
 Enter your choice9

...Displaying List From End using Recursion...

EMPTY LIST::No nodes to display
10101 1111
YOU WANT TO CONTINUE (Y/N)y
 ------------------------------
 Operations on singly linked list
 ------------------------------
 1.Insert node at first
2.Insert node at last
3.Insert node at position
4.Sorted Linked List in Ascending Order
5.Delete Node from any Position
6.Update Node Value
7.Search Element in the linked list
8.Display List from Beginning to end
9.Display List from end using Recursion
10.Exit
~~~~~~~~~~~~~~~~~~~~~~~~~~~~~~~~~~
 Enter your choice10
 ...Exiting...
```

4.9. C Structure Using Typedef

Using typedef with Structures

- It allows us to introduce synonyms for data types which could have been declared some other way.
- It is used to give New name to the Structure.
- New name is used for Creating instances, Passing values to function,declaration etc…

Example of typedef Sample Program

```
#include<stdio.h>

int main()
```

```
{
typedef int Number;
Number num1 = 40,num2 = 20;
Number answer;

answer = num1 + num2;
printf("Answer : %d",answer);
return(0);
}
```

Output

```
Answer: 60
```

Explanation of Program

In the above program we have used typedef to create alias name to data type. We have created alias name to _int' data type. We have given new name to integer data type i.e _Number'.

Syntax to Use typedef with Structures

```
typedef int XYZ;
 XYZ age;
 XYZ roll;
```

Different Ways of Declaring Structure using typedef

```
typedef struct
{
 char ename[30];
 int ssn;
 int deptno;
}employee;
```

Or

```
typedef struct Record
 {
 char ename[30];
 int ssn;
 int deptno;
 }employee;
```

In the second example, Record is tag-name. 'employee' is nothing but New Data Type. We can now create the variables of type 'employee' Tag name is optional.

Declaring Variable

```
employee e1,e2;
```

Example of Using Typedef For Declaring Structure

```
#include<stdio.h>

typedef struct b1 {
 char bname[30];
 int ssn;
 int pages;
}book;

book b1 = {"Let Us C",1000,90};

int main()
{
printf("\nName of Book : %s",b1.bname);
printf("\nSSN of Book : %d",b1.ssn);
printf("\nPages in Book : %d",b1.pages);
return(0);
}
```

Output

```
Name of Book: Let Us C
SSNof Book: 1000
Pages in Book : 90
```

UNIT V
FILE PROCESSING

Files – Types of file processing: Sequential access, Random access – Sequential access file -
Example Program: Finding average of numbers stored in sequential access file - Random access file -Example Program: Transaction processing using random access files – Command line arguments

5. INTRODCUTION

5.1. What is File?

A file represents a sequence of bytes on the disk where a group of related data is stored. File is created for permanent storage of data. It is a readymade structure. In C programming, file is a place on your physical disk where information is stored.

Why files are needed?

- When a program is terminated, the entire data is lost. Storing in a file will preserve your data even if the program terminates.
- If you have to enter a large number of data, it will take a lot of time to enter them all. However, if you have a file containing all the data, you can easily access the contents of the file using few commands in C.
- You can easily move your data from one computer to another without any changes.

5.2. Types of Files

When dealing with files, there are two types of files you should know about:

1. Text files
2. Binary files

Text files

- Text files are the normal .txt files that you can easily create using Notepad or any simple text editors.
- When you open those files, you'll see all the contents within the file as plain text. You can easily edit or delete the contents.
- They take minimum effort to maintain, are easily readable, and provide least security and takes bigger storage space.

Binary files

- Binary files are mostly the .bin files in your computer.
- Instead of storing data in plain text, they store it in the binary form (0's and 1's).
- They can hold higher amount of data, are not readable easily and provides a better security than text files.

File Operations

In C, you can perform four major operations on the file, either text or binary:

1. Creating a new file
2. Opening an existing file
3. Closing a file
4. Reading from and writing information to a file

Working with files

When working with files, you need to declare a pointer of type file. This declaration is needed for communication between the file and program.

*FILE *fptr;*

Opening a file - for creation and edit

Opening a file is performed using the library function in the "stdio.h" header file: fopen().

The syntax for opening a file in standard I/O is:

ptr = fopen("fileopen","mode") **For Example**
fopen("E:\\cprogram\\newprogram.txt","w");

fopen("E:\\cprogram\\oldprogram.bin","rb");

- Let's suppose the file newprogram.txt doesn't exist in the location **E:\cprogram**. The first function creates a new file named newprogram.txt and opens it for writing as per the mode 'w'.
- The writing mode allows you to create and edit (overwrite) the contents of the file.
- Now let's suppose the second binary file oldprogram.bin exists in the location **E:\cprogram**. The second function opens the existing file for reading in binary mode 'rb'.
- The reading mode only allows you to read the file, you cannot write into the file.

Opening Modes in Standard I/O		
File Mode	**Meaning of Mode**	**During Inexistence of file**
r	Open for reading.	If the file does not exist, fopen() returns NULL.
rb	Open for reading in binary mode.	If the file does not exist, fopen() returns NULL.
w	Open for writing.	If the file exists, its contents are overwritten. If the file does not exist, it will be created.
wb	Open for writing in binary mode.	If the file exists, its contents are overwritten. If the file does not exist, it will be created.
A	Open for append. i.e, Data is added to end of file.	If the file does not exists, it will be created.
ab	Open for append in binary mode. i.e, Data is added to end of file.	If the file does not exists, it will be created.
r+	Open for both reading	If the file does not exist,

	and writing.	fopen() returns NULL.
rb+	Open for both reading and writing in binary mode.	If the file does not exist, fopen() returns NULL.
w+	Open for both reading and writing.	If the file exists, its contents are overwritten. If the file does not exist, it will be created.
wb+	Open for both reading and writing in binary mode.	If the file exists, its contents are overwritten. If the file does not exist, it will be created.
a+	Open for both reading and appending.	If the file does not exists, it will be created.
ab+	Open for both reading and appending in binary mode.	If the file does not exists, it will be created.

Closing a File

The file (both text and binary) should be closed after reading/writing.

Closing a file is performed using library function fclose().

fclose(fptr); //fptr is the file pointer associated with file to be closed.

Reading and writing to a text file

For reading and writing to a text file, we use the functions **fprintf()** and ***fscanf().***

They are just the file versions of ***printf()*** and ***scanf().*** The only difference is that, ***fprint*** and ***fscanf*** expects a pointer to the structure FILE.

Writing to a text file

Example of Write to a text file using fprintf()

```
#include <stdio.h>
int main()
{
 int num;
 FILE *fptr;
 fptr = fopen("C:\\program.txt","w");
```

```
if(fptr == NULL)
{
printf("Error!");
exit(1);
}

printf("Enter num: ");
scanf("%d",&num);

fprintf(fptr,"%d",num);
fclose(fptr);

return 0;
}
```

This program takes a number from user and stores in the file ***program.txt.***

After you compile and run this program, you can see a text file ***program.txt*** created in C drive of your computer. When you open the file, you can see the integer you entered.

5.2.1. Reading from a text file

Example of Read from a text file using fscanf()

```
#include <stdio.h>
int main()
{
int num;
FILE *fptr;

if ((fptr = fopen("C:\\program.txt","r")) == NULL){
printf("Error! opening file");

// Program exits if the file pointer returns NULL.
exit(1);
}
fscanf(fptr,"%d", &num);

printf("Value of n=%d", num);
fclose(fptr);
```

```
return 0;
}
```

This program reads the integer present in the ***program.txt*** file and prints it onto the screen.

If you successfully created the file from **Example 1**, running this program will get you the integer you entered.

Other functions like ***fgetchar(), fputc()*** etc. can be used in similar way.

5.2.2. Reading and Writing to a Binary File

Functions ***fread()*** and ***fwrite()*** are used for reading from and writing to a file on the disk respectively in case of binary files.

Writing to a binary file

To write into a binary file, you need to use the function ***fwrite()***. The functions takes four arguments: Address of data to be written in disk, Size of data to be written in disk, number of such type of data and pointer to the file where you want to write.

fwrite(address_data,size_data,numbers_data,pointer_to_file);

Example 3: Writing to a binary file using fwrite()

```
#include <stdio.h>
struct threeNum
{
 int n1, n2, n3;
};
int main()
{
 int n;
 struct threeNum num;
 FILE *fptr;
 if ((fptr = fopen("C:\\program.bin","wb")) == NULL){
 printf("Error! opening file");

 // Program exits if the file pointer returns NULL.
 exit(1);
 }
```

```
for(n = 1; n < 5; ++n)
{
num.n1 = n;
num.n2 = 5n;
num.n3 = 5n + 1;
fwrite(&num, sizeof(struct threeNum), 1, fptr);
}
fclose(fptr);
return 0;
}
```

In this program, you create a new file program.bin in the C drive.

We declare a structure three Num with three numbers - ***n1, n2 and n3,*** and define it in the main function as num.
Now, inside the for loop, we store the value into the file using ***fwrite.***

The first parameter takes the address of *num* and the second parameter takes the size of the structure threeNum.

Since, we're only inserting one instance of *num*, the third parameter is 1. And, the last parameter **fptr* points to the file we're storing the data.
Finally, we close the file.

Reading from a binary file

Function **fread()** also take 4 arguments similar to *fwrite()* function as above.
fread(address_data,size_data,numbers_data,pointer_to_file);

Example of Reading from a binary file using fread()

```
#include <stdio.h>
struct threeNum
{
int n1, n2, n3;
};

int main()
```

```
{
 int n;
 struct threeNum num;
 FILE *fptr;

 if ((fptr = fopen("C:\\program.bin","rb")) == NULL){
 printf("Error! opening file");

// Program exits if the file pointer returns NULL.
 exit(1);
 }

 for(n = 1; n < 5; ++n)
 {
fread(&num, sizeof(struct threeNum), 1, fptr);
printf("n1: %d\tn2: %d\tn3: %d", num.n1, num.n2, num.n3);
 }
 fclose(fptr);
 return 0;
}
```

In this program, you read the same file program.bin and loop through the records one by one.

In simple terms, you read one threeNum record of threeNum size from the file pointed by ***fptr** into the structure ***num.***

You'll get the same records you inserted in Example 3.

Getting data using fseek()

If you have many records inside a file and need to access a record at a specific position, you need to loop through all the records before it to get the record.

This will waste a lot of memory and operation time. An easier way to get to the required data can be achieved using ***fseek().***

As the name suggests, ***fseek()*** seeks the cursor to the given record in the file.

Syntax of fseek()

fseek(FILE * stream, long int offset, int whence)

The first parameter stream is the pointer to the file. The second parameter is the position of the record to be found, and the third parameter specifies the location where the offset starts.

Different Whence in fseek	
Whence	**Meaning**
SEKK_SET	Starts the offset from the beginning of the file.
SEKK_END	Starts the offset from the end of the file.
SEKK_CUR	Starts the offset from the current location of the cursor in the file.

Example of fseek()

```
#include <stdio.h>
struct threeNum
{
 int n1, n2, n3;
};
int main()
{
 int n;
 struct threeNum num;
 FILE *fptr;

 if ((fptr = fopen("C:\\program.bin","rb")) == NULL){
 printf("Error! opening file");

 // Program exits if the file pointer returns NULL.
 exit(1);
 }

 // Moves the cursor to the end of the file
 fseek(fptr, sizeof(struct threeNum), SEEK_END);

 for(n = 1; n < 5; ++n)
 {
fread(&num, sizeof(struct threeNum), 1, fptr);
printf("n1: %d\tn2: %d\tn3: %d", num.n1, num.n2, num.n3);
```

```
}
fclose(fptr);
return 0;
}
```

This program will start reading the records from the file program.bin in the reverse order (last to first) and prints it.

5.3. Sequential Access And Random Access Files

There are two main ways a file can be organized:

1. **Sequential Access** — The data are placed in the file in a sequence like beads on a string. Data are processed in sequence, one after another. To reach a particular item of data, all the data that proceeds it first must be read.
2. **Random Access** — The data are placed into the file by going directly to the location in the file assigned to each data item. Data are processed in any order. A particular item of data can be reached by going directly to it, without looking at any other data.

A sequential file works like a reel of tape. (In fact, sequential files are often stored on reels of magnetic tape.) Data in a sequential file are processed in order, starting with the first item, the processing the second, then the third and so on. To reach data in the middle of the file you must go through all the data that preceeds it.

A random access file works like a sampler box of chocolate. You can access a particular item by going directly to it. Data in a random acess file may be accessed in any order. To read data in the middle of the file you can go directly to it. Random access files are sometimes called **direct access** files.

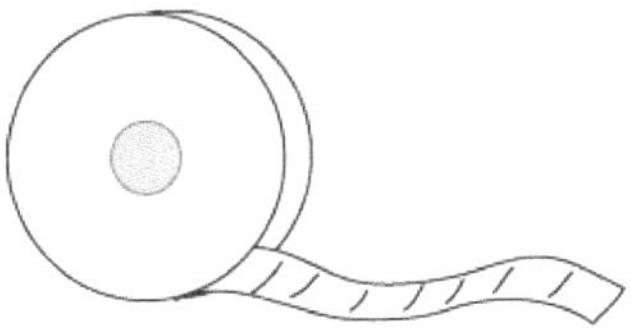

Sequential File

Random Access File

You might think that all files should be random access. But random access files are much more difficult to implement and to process. They are expensive in computer time and resources. However, they are convenient for the users of a program. For example, a corporation can easily answer telephone queries from a customer if it can go directly to the customer's records.

5.3.1. Sequential Access

A Simple C Program to open, read and close the file

```
#include <stdio.h>
int main()
{
/* Pointer to the file */
FILE *fp1;
/* Character variable to read the content of file */
char c;

/* Opening a file in r mode*/
fp1= fopen ("C:\\myfiles\\newfile.txt", "r");

/* Infinite loop –I have used break to come out of the loop*/
while(1)
{
c = fgetc(fp1);
if(c==EOF)
break;
else
printf("%c", c);
}
fclose(fp1);
return 0;
}
```

In the above program, we are opening a file **newfile.txt** in **r** mode, reading the content of the file and displaying it on the console. lets understand the each operation in detail:

Opening a file

fopen() function is used for opening a file.

Syntax

FILE pointer_name = fopen ("file_name", "Mode");

pointer_name can be anything of your choice.

file_name is the name of the file, which you want to open. Specify the full path here like **"C:\\myfiles\\newfile.txt".**

While opening a file, you need to specify the mode. The mode that we use to read a file is —r‖ which is —read only mode‖.

For example

*FILE*fp;*

fp = fopen("C:\\myfiles\\newfile.txt", "r");

- The address of the first character is stored in pointer fp.
- How to check whether the file has opened successfully?
- If file does not open successfully then the pointer will be assigned a NULL value, so you can write the logic like this:
- This code will check whether the file has opened successfully or not. If the file does not open, this will display an error message to the user.

```
..
FILE fpr;
fpr = fopen("C:\\myfiles\\newfile.txt", "r");
if (fpr == NULL)
{
puts("Error while opening file");
exit();
}
```

Various File Opening Modes

The file is opened using fopen() function, while opening you can use any of the following mode as per the requirement.

Mode "r": It is a read only mode, which means if the file is opened in r mode, it won't allow you to write and modify content of it. When fopen() opens a file successfully then it returns the address of first character of the file, otherwise it returns NULL.

Mode "w": It is a write only mode. The fopen() function creates a new file when the specified file doesn't exist and if it fails to open file then it returns NULL.

Mode "a": Using this mode Content can be appended at the end of an existing file. Like Mode —w‖, fopen() creates a new file if it file doesn't exist. On unsuccessful open it returns NULL.

File Pointer points to: last character of the file.

Mode "r+": This mode is same as mode —r‖; however you can perform various operations on the file opened in this mode. You are allowed to read, write and modify the content of file opened in —r+‖ mode.

File Pointer points to: First character of the file.

Mode —w+‖: Same as mode —w‖ apart from operations, which can be performed; the file can be read, write and modified in this mode.

Mode —a+‖: Same as mode —a‖; you can read and append the data in the file, however content modification is not allowed in this mode.

Reading a File

To read the file, we must open it first using any of the mode, for example if you only want to read the file then open it in —r‖ mode. Based on the mode selected during file opening, we are allowed to perform certain operations on the file.

C Program to read a file

fgetc( ): This function reads the character from current pointer's position and upon successful read moves the pointer to next character in the file. Once the pointers reaches to the end of the file, this function returns EOF (End of File). We have used EOF in our program to determine the end of the file.

```
#include <stdio.h>
int main()
```

```
{
/* Pointer to the file */
FILE *fp1;
/* Character variable to read the content of file */
char c;

/* Opening a file in r mode*/
fp1= fopen ("C:\\myfiles\\newfile.txt", "r");

/* Infinite loop –I have used break to come out of the loop*/
while(1)
{
c = fgetc(fp1);
if(c==EOF)
break;
else
printf("%c", c);
}
fclose(fp1);
return 0;
}
```

Writing to a file

To write the file, we must open the file in a mode that supports writing. For example, if you open a file in –r‖ mode, you won't be able to write the file as —r‖ is read only mode that only allows reading.

Example of C Program to write the file

This program asks the user to enter a character and writes that character at the end of the file. If the file doesn't exist then this program will create a file with the specified name and writes the input character into the file.

```
#include <stdio.h>
int main()
{
char ch;
FILE *fpw;
```

```
fpw = fopen("C:\\newfile.txt","w");

if(fpw == NULL)
{
printf("Error");
exit(1);
}

printf("Enter any character: ");
scanf("%c",&ch);

/* You can also use fputc(ch, fpw);*/
fprintf(fpw,"%c",ch);
fclose(fpw);

return 0;
}
```

Closing a file
fclose(fp);

The fclose( ) function is used for closing an opened file. As an argument you must provide a pointer to the file that you want to close.

An example to show Open, read, write and close operation in C

```
#include <stdio.h>
int main()
{
char ch;

/* Pointer for both the file*/
FILE *fpr, *fpw;
/* Opening file FILE1.C in "r" mode for reading */
fpr = fopen("C:\\file1.txt", "r");

/* Ensure FILE1.C opened successfully*/
if (fpr == NULL)
{
```

```
puts("Input file cannot be opened");
}

/* Opening file FILE2.C in "w" mode for writing*/
fpw= fopen("C:\ \file2.txt", "w");

/* Ensure FILE2.C opened successfully*/
if (fpw == NULL)
{
puts("Output file cannot be opened");
}

/*Read & Write Logic*/
while(1)
{
ch = fgetc(fpr);
if (ch==EOF)
break;
else
fputc(ch, fpw);
}

/* Closing both the files */
fclose(fpr);
fclose(fpw);

return 0;
}
```

How to read/ write (I/O) Strings in Files – fgets & fputs

Here we will discuss how to read and write strings into a file.

char *fgets(char *s, int rec_len, FILE *fpr)

s: Array of characters to store strings.

rec_len: Length of the input record.

fpr: Pointer to the input file.

5.4. Example Program

5.4.1. Example to read the strings from a file in C programming

```
#include <stdio.h>
int main()
{
FILE *fpr;
/*Char array to store string */
char str[100];
/*Opening the file in "r" mode*/
fpr = fopen("C:\\mynewtextfile.txt", "r");

/*Error handling for file open*/
if (fpr == NULL)
{
 puts("Issue in opening the input file");
}

/*Loop for reading the file till end*/
while(1)
{
 if(fgets(str, 10, fpr) ==NULL)
break;
 else
printf("%s", str);
}
/*Closing the input file after reading*/
fclose(fpr);
return 0;
}
```

In the above example we have used fgets function like this:

```
fgets(str, 10, fpr)
```

Here str represents the string (array of char) in which you are storing the string after reading it from file.

- 10 is the length of the string that needs to be read every time.
- fpr is pointer to file, which is going to be read.

Why I used if(fgets(str, 10, fpr)==NULL as a logic to determine end of the file?

In the above examples, we have used ch==EOF to get to know the end of the file. Here we have used this logic beause fgets returns NULL when there is no more records are available to be read.

5.4.2. C Program – Writing string to a file

```
int fputs ( const char * s, FILE * fpw );
char *s – Array of char.
FILE *fpw – Pointer (of FILE type) to the file, which is going to be
written.

#include <stdio.h>
int main()
{
 FILE *fpw;

 /*Char array to store strings */
 char str[100];

 /*Opening the file FILEW.TXT in "w" mode for writing*/
 fpw = fopen("C:\\mynewtextfile2.txt", "w");

 /*Error handling for output file*/
 if (fpw== NULL)
 {
puts("Issue in opening the Output file");
 }

 printf("Enter your string:");

 /*Stored the input string into array – str*/
 gets(str);

 /* Copied the content of str into file –
 * mynewtextfile2.txt using pointer – fpw
```

```
*/
fputs(str, fpw);

/*Closing the Output file after successful writing*/
fclose(fpw);
return 0;
}
```

fputs takes two arguments
fputs(str, fpw)

- str – str represents the array, in which string is stored.
- fpw – FILE pointer to the output file, in which record needs to be written.

Point to note about fputs

fputs by default doesn't add new line after writing each ecord, in order to do that manually – you can have the following statement after each write to the file.

fputs("\n", fpw);

C FILE I/O for Binary files

So far, we have learned file operations on text files, what if the files are binary (such as .exe file). The above programs will not work for binary files, however there is a minor change in handling Binary files. The main difference is the file name & modes. Lets understand this with the help of an example. Lets say I have two binary files bin1.exe & bin2.exe – I want to copy content of bin1.exe to bin2.exe:

Example of Reading and Writing Binary Files in C

```
#include <stdio.h>
int main()
{
char ch;

/* Pointers for both binary files*/
FILE *fpbr, *fpbw;
```

```
/* Open for bin1.exe file in rb mode */
fpbr = fopen("bin1.exe", "rb");

/* test logic for successful open*/
if (fpbr == NULL)
{
puts("Input Binary file is having issues while opening");
}

/* Opening file bin2.exe in "wb" mode for writing*/
fpbw= fopen("bin2.exe", "wb");

/* Ensure bin2.exe opened successfully*/
if (fpbw == NULL)
{
 puts("Output binary file is having issues while opening");
}

/*Read & Write Logic for binary files*/
while(1)
{
ch = fgetc(fpbr);
if (ch==EOF)
break;
 else
 fputc(ch, fpbw);
 }

 /* Closing both the binary files */
 fclose(fpbr);
 fclose(fpbw);

 return 0;
}
```

Note: File opening modes are –rb‖ and —wb‖ instead of –r‖ & –w‖..

5.4.3. Example program of finding average of numbers stored in sequential access file

Data processing is an important application in computing. To process large amounts of data, we need some way to store and access that data. The most natural way to accomplish this is to store data in files and then provide some mechanism for reading and writing data from and to those files.

C offers some very simple mechanisms for reading and writing data from files. The simplest of these mechanisms is the mechanism for working with data stored in text files. C also offers ways to work with binary files, but since text files will suffice for our needs in this course I will only be discussing how to work with text files.

Basics of files

The following example is a program that opens up a text file named "data.txt" containing a list of integers, reads thoses integers, and computes their sum.

```
#include <stdio.h>
int main (int argc, const char * argv[])
{
FILE *input;
int term, sum;

sum = 0;
input = fopen("data.txt","r");
while(!feof(input))
{
fscanf(input,"%d",&term);
sum = sum + term;
}
fclose(input);

printf("The sum of the numbers is %d.\n",sum);

return 0;
}
```

The first step in working with a file is to create a *file pointer variable.* This is done with a variable declaration

FILE *input;

The FILE data type is a special data type defined in the stdio.h header file. Most functions designed to work with FILE variables will expect us to pass a *pointer* to that variable. The * you see in the variable declaration indicates that the variable input is a pointer to a FILE. We will discuss pointer variables in detail later in the term. For now, all you have to know is that the line shown above is the appropriate syntax for setting up a FILE pointer variable, and that will suffice for the time being.

The second step in working with a file is to open the file for reading or writing. This is done by calling the fopen function.

input = fopen("data.txt","r");

fopen takes two parameters. The first is the name of the file to open, and the second is a text string that specifies an *open mode* to use. To open a text file for reading data, we use an open mode of "r". To open a text file for writing, we will use an open mode of "w".

In specifying the name of the file to open we can use two methods. The simple method is to just name the file. For this to work correctly, the file has to be placed in a location where it can be found easily. In the case of Visual Studio, this means placing the text file in the same folder as the source code for the program. In other development environments this means placing the file in the same directory where the compiled program resides after compilation. You may have to experiment a little with your development environment to figure out where this default location is. A second, more reliable alternative for specifying a file is to give a complete path to the file. On Windows a complete path can look something like the following:

C:\Users\Joe Gregg\Documents\CMSC 110\Examples\AddFile\AddFile\data.txt

Since the backslash (\) character that Windows uses for file paths has a special meaning in C text strings, to encode a backslash in a string in C you will have to use \\ in place of each \ in the path. This looks something like the following:

input = fopen("C:\\Users\\Joe Gregg\\Documents\\CMSC 110\\Examples\\AddFile\\AddFile\\data.txt","r");

On Unix based operating systems such as Mac OS and Linux, file paths use forward slashes instead of backslashes. This makes specifying a file by its complete file path easier to do.

input = fopen("Mac HD/Users/Joe Gregg/Documents/CMSC 110/Examples/AddFile/data.txt","r");

Once the file is open for reading we can read data from it. Reading data from a text file is very similar to reading data from the console. Instead of using scanf(), we use fscanf().

fscanf(input,"%d",&term);

fscanf takes three parameters: the file pointer variable, a format specifier string, and a list of things to read.

Each time you read a data item from an open file you advance through the file. After performing a long enough sequence of reads you will have read to the end of the file. To tell that you have reached the end of the file, you can call feof(input). This function returns 0 if we have not reached the end of the file and 1 if we have. Most commonly, feof() is used to control a loop that reads data items from the file until we have reached the end of the file.

```
while(!feof(input))
{
fscanf(input,"%d",&term);
sum += term;
}
```

Once you have reached the end of the file you should close it. This is done by calling the fclose() function.

fclose(input);

If for some reason you need to scan through the contents of a file more than once, you will have to close and reopen the file for each subsequent pass through the file.

Writing to files

Writing data to text files is very similar to reading from text files. To write output to a file we use fprintf() in place of the usual printf() used to print output to the console.

Here is an example program that generates a sequence of 1000 pseudo-random integers in the range from 0 to 9 and writes them to a text file, one number per line.

The program generates this sequence of numbers by using a *linear congruential random number generator.* This is a simple mathematical formula that generates a sequence of integers. To form the sequence we pick an arbitrary integer as the *seed* for the sequence. We then substitute that starting value into a simple linear function to generate the next number in the sequence. Most often that simple linear function uses two large prime numbers as its coefficients.

$x_1 = 12553\, x_0 + 15401$

We repeat that process to make a whole sequence of numbers. If we did just that, the sequence would typically march off toward positive or negative infinity. To prevent that, we mod each number by a large prime integer b on each round.

$x_1 = (12553\, x_0 + 15401)\ mod\ 6133$

This guarantees that the numbers stay trapped in the range from 0 to b-1. Often, the sequence of integers generated by this process is not immediately useful to us, because, say, we may be interested in generating a sequence of numbers in the range from 0 to c-1 for some other integer c. That is easy to fix. If $c < b$, we can compute x % c for each number we generate and save the list of x % c values as we generate them.

Here now is the program that does all this.

```
#include <stdio.h>

int main (int argc, const char * argv[])
{
FILE *output;
int x, n;

// Seed value to start the sequence
x = 152;
n = 0;
```

```
output = fopen("random.txt","w");

while(n < 1000)
{
// Generate the next number in the sequence
// Mod by 11657 to keep the numbers in a
// modest range.
x = (17389*x+12553)%11657;
n = n + 1;
// Save a number to the output file.
// Since we want the output file to contain
// numbers in the range from 0 to 9 we
// have to mod x by 10 to get the desired
// value.
fprintf(output,"%d\n",x%10);
} fclose(output);

printf("Done.\n");

return 0;
}
```

Some comments about this program:

1. To open the output file for writing we use fopen as usual, but with a mode of "w" for writing. This will create the file if it does not already exist.
2. If the file exists when we open it, its contents will get replaced by what we are about to write. Alternatively, we could use a mode specifier of "a", which means "append." This will append the data we are writing to what is there already.
3. To send output to the file we use fprintf() instead of the usual printf(). Note that if we want numbers to appear one per line we have to explicitly write the '\n' characters, much as we would when working with printf().

4. When we are done with the file we have to remember to close it with fclose(). Forgetting to close the file can result in some circumstances in some or all of the data not getting written to the file.

An example of data processing

The next example program carries out a simple data processing task on some data stored in a text file. The task is to compute and report some simple statistics on the data in a file. We will read a sequence of real numbers from a file and compute the average and standard deviation of the numbers in the list.

The standard deviation of a sequence of *N* numbers x_n is

$$\sigma = \sqrt{\frac{\sum_{n=1}^{N}(x_n - \overline{x})^2}{N}}$$

where *x* is the average of the numbers in the list.

The example program below computes the standard deviation for a list of numbers stored in a file named "data.txt." The program makes two passes through the file. The first pass is needed to compute the average of the numbers in the file: that work gets done in the average() function. The second pass uses that average to compute a *deviation* (x_n - *x*) for each number in the list. The function stddev() computes the sum of the squares of those deviations, averages them, and then computes to the square root of that average to obtain the standard deviation.

```
#include <stdio.h>
#include <math.h>

// Read the numbers in the indicated
// file and return their average
float average(FILE *input) {
float term,sum;
int n;

sum = 0.0;
n = 0;
```

```
while(!feof(input))
{
fscanf(input,"%f",&term);
sum = sum + term;
n = n + 1;
}

return sum/n;
}

// Compute the standard deviation of the
// numbers in the file given the mean.
float stddev(FILE *input,float mean) {
float term,delta,total;
int n;

total = 0.0;
n = 0;

while(!feof(input))
{
fscanf(input,"%f",&term);
delta = term - mean;
total = total + delta*delta;
n = n + 1;
}

return sqrt(total/n);
}

int main (int argc, const char * argv[])
{
FILE *input;
float mean,sdev;

input = fopen("data.txt","r");
mean = average(input);
```

```
fclose(input);

input = fopen("data.txt","r");
sdev = stddev(input,mean);
fclose(input);

printf("The average of the numbers is %f.\n",mean);
printf("The standard deviation of the numbers is %f.\n",sdev);
}
```

Note that since we have to make two passes through the data file we will have to open and close the file twice. That happens in main(). Once the file is open and prepared for a reading pass we can pass the FILE pointer variable to a function for it to use to access the file.

5.5. Random Access to File

Individual records of a random-access file are normally fixed in length and may be accessed directly (and thus quickly) without searching through other records.

This makes random-access files appropriate for:

- Airline reservation systems
- Banking systems
- Point-of-sale systems
- Other kinds of transaction-processing systems that require rapid access to specific data.

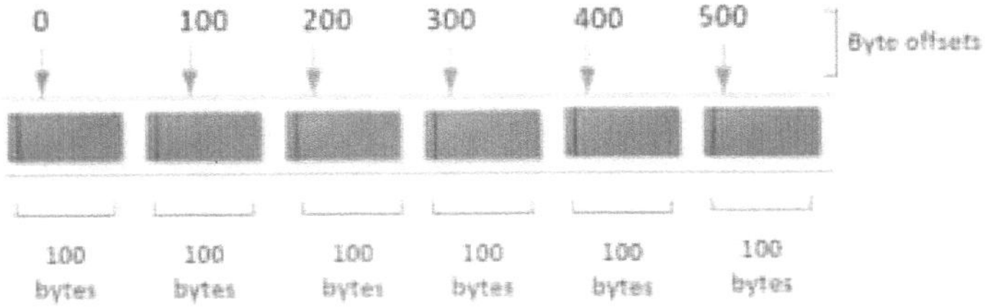

Random-Access File

Fixed-length records enable data to be inserted in a random-access file without destroying other data in the file. Data stored previously can also be updated or deleted without rewriting the entire file.

5.5.1. Creating a Random-Access File

Functions fwrite and fread are capable of reading and writing arrays of data to and from disk. The third argument of both fread and fwrite is the number of elements in the array that should be read from or written to disk.

The following program open a random-access file, define a record format using a struct, write data to the disk and close the file.

```
// Creating a random-access file sequentially
#include <stdio.h>
// clientData structure definition
struct clientData {
unsigned int acctNum; // account number
char lastName[ 15 ]; // account last name
char firstName[ 10 ]; // account first name
double balance; // account balance
}; // end structure clientData
int main( void ) {
unsigned int i; // counter used to count from 1-100
// create clientData with default information
struct clientData blankClient = { 0, "", "", 0.0 };
FILE *cfPtr; // credit.dat file pointer
// fopen opens the file; exits if file cannot be opened
if ( ( cfPtr = fopen( "credit.dat", "wb" ) ) == NULL ) {
puts( "File could not be opened." );
} // end if
else {
// output 100 blank records to file
for ( i = 1; i <= 100; ++i ) {
fwrite( &blankClient, sizeof( struct clientData ), 1, cfPtr );
} // end for
fclose ( cfPtr ); // fclose closes the file
} // end else
} // end main
```

5.5.2. Writing Data Randomly to a Random-Access File

The program writes data to the file "credit.dat". It uses the combination of fseek and fwrite to store data at specific locations in the file.

Function fseek sets the file position pointer to a specific position in the file, then fwrite writes the data.

```
// Writing data randomly to a random-access file
#include <stdio.h>
// clientData structure definition
struct clientData {
unsigned int acctNum; // account number
char lastName[ 15 ]; // account last name
char firstName[ 10 ]; // account first name
double balance; // account balance
}; // end structure clientData

int main( void ){
FILE *cfPtr; // credit.dat file pointer
// create clientData with default information
struct clientData client = { 0, "", "", 0.0 };
// fopen opens the file; exits if file cannot be opened
if ( ( cfPtr = fopen( "credit.dat", "rb+" ) ) == NULL ) {
puts( "File could not be opened." );
} // end if
else {
// require user to specify account number
printf( "%s", "Enter account number"
 " ( 1 to 100, 0 to end input )\n? " );
scanf( "%d", &client.acctNum );
// user enters information, which is copied into file
while ( client.acctNum != 0 ) {
// user enters last name, first name and balance
printf( "%s", "Enter lastname, firstname, balance\n? " );
// set record lastName, firstName and balance value
fscanf( stdin, "%14s%9s%lf", client.lastName,
client.firstName, &client.balance );
// seek position in file to user-specified record
fseek( cfPtr, ( client.acctNum - 1 ) *
```

```
sizeof( struct clientData ), SEEK_SET );
// write user-specified information in file
fwrite( &client, sizeof( struct clientData ), 1, cfPtr );
// enable user to input another account number
printf( "%s", "Enter account number\n? " );
scanf( "%d", &client.acctNum );
} // end while
fclose( cfPtr ); // fclose closes the file
} // end else
} // end main
```

Output

```
Enter account number ( 1 to 100, 0 to end input )
? 37
Enter lastname, firstname, balance
? Barker Doug 0.00
Enter account number
? 29
Enter lastname, firstname, balance
? Brown Nancy -24.54
Enter account number
? 96
Enter account number
? 0
```

There is no need to read each record sequentially, if we want to access a particular record.C supports these functions for random access file processing.

- fseek()
- ftell()
- rewind()

fseek():

This function is used for seeking the pointer position in the file at the specified byte.

Syntax: fseek(file pointer, displacement, pointer position);

Where

file pointer ---- It is the pointer which points to the file.
displacement ---- It is positive or negative.This is the number of bytes which are skipped backward (if negative) or forward(if positive) from the current position.This is attached with L because this is a long integer.

Pointer position

This sets the pointer position in the file.

Value	Pointer Position
0	Beginning of file.
1	Current position
2	End of file

Example

1. **fseek(p,10L,0)**
   0 means pointer position is on beginning of the file,from this statement pointer position is skipped 10 bytes from the beginning of the file.
2. **fseek(p,5L,1)**
   1 means current position of the pointer position.From this statement pointer position is skipped 5 bytes forward from the current position.
3. **fseek(p,-5L,1)**
   From this statement pointer position is skipped 5 bytes backward from the current position.

ftell()

This function returns the value of the current pointer position in the file. The value is count from the beginning of the file.
Syntax: ftell(fptr);
Where fptr is a file pointer.

Example: ftell()

The following example shows the usage of ftell() function.

```
#include <stdio.h>

int main () {
 FILE *fp;
 int len;
```

```
fp = fopen("file.txt", "r");
if( fp == NULL ){
perror ("Error opening file");
return(-1);
}
fseek(fp, 0, SEEK_END);

len = ftell(fp);
fclose(fp);

printf("Total size of file.txt = %d bytes\n", len);

return(0);
}
```

Let us assume we have a text file **file.txt**, which has the following content –

```
This is studentspoint.com
```

Now let us compile and run the above program that will produce the following result if file has above mentioned content otherwise it will give different result based on the file content –

```
Total size of file.txt = 27 bytes
```

rewind()

This function is used to move the file pointer to the beginning of the given file.

Syntax: rewind(fptr);

Where fptr is a file pointer.

Example: rewind()

The following example shows the usage of rewind() function.

```
#include <stdio.h>


int main () {
char str[] = "This is tutorialspoint.com";
FILE *fp;
int ch;

/* First let's write some content in the file */
fp = fopen( "file.txt" , "w" );
```

```
fwrite(str , 1 , sizeof(str) , fp );
fclose(fp);

fp = fopen( "file.txt" , "r" );
while(1) {
ch = fgetc(fp);
if( feof(fp) ) {
break ;
}
printf("%c", ch);
} rewind(fp);
printf("\n");
while(1) {
ch = fgetc(fp);
if( feof(fp) ) {
break ;
}
printf("%c", ch);

}
fclose(fp);

return(0);
}
```

Let us assume we have a text file **file.txt** that have the following content –

This is studentspoint.com

Now let us compile and run the above program to produce the following result –

This is studentspoint.com

This is studentspoint.com

Example program for fseek():

Write a program to read last _n‘ characters of the file using appropriate file functions (Here we need fseek() and fgetc()).

```
#include<stdio.h>
#include<conio.h>
void main()
{
FILE *fp;
char ch;
clrscr();
fp=fopen("file1.c", "r");
if(fp==NULL)
printf("file cannot be opened");
else
{
 printf("Enter value of nto read last „n" characters");
scanf("%d",&n); fseek(fp,-
n,2);
while((ch=fgetc(fp))!=EOF)
{
printf("%c\t",ch);
}
}
fclose(fp);
getch();
}
```

Output

```
It depends on the content in the file.
```

5.6. Case Study: A Transaction Processing Program

- **This program**
  - Demonstrates using random access files to achieve instant access processing of a bank's account information
- **We will**
  - Update existing accounts
  - Add new accounts
  - Delete accounts
  - Store a formatted listing of all accounts in a text file

/*This program reads a random access file sequentially, updates data already written to the file, creates new data to be placed in the file, and deletes data already in the file. */

- Define struct
  1. Function prototypes
  2. Initialize variables
  3. Link pointer and open file
- Input choice
  1. Perform action
  2. Close file
- Function definitions
- Program Output

```
#include <stdio.h>
struct clientData {
int acctNum;
char lastName[ 15 ];

char firstName[ 10 ];
double balance;
};
int enterChoice( void );
void textFile( FILE * );
void updateRecord( FILE * );
void newRecord( FILE * );
void deleteRecord( FILE * );

int main()
{
FILE *cfPtr;
int choice;

if ( ( cfPtr = fopen( "credit.dat", "r+" ) ) == NULL )
printf( "File could not be opened.\n" );

else {
while ( ( choice = enterChoice() ) != 5 ) {
switch ( choice ) {
```

```
case 1:
textFile( cfPtr );
break;

case 2:
updateRecord( cfPtr );
break;

case 3:
newRecord( cfPtr );
break;

case 4:
deleteRecord( cfPtr );
break;

}
}
fclose( cfPtr );
}
return 0;
}
void textFile( FILE *readPtr )
{
FILE *writePtr;
struct clientData client = { 0, "", "", 0.0 };
if ( ( writePtr = fopen( "accounts.txt", "w" ) ) == NULL )
printf( "File could not be opened.\n" );
else {
rewind( readPtr );
fprintf( writePtr, "%-6s%-16s%-11s%10s\n",
"Acct", "Last Name", "First Name","Balance" );
while ( !feof( readPtr ) )
fread( &client, sizeof( struct clientData ), 1, readPtr );
if ( client.acctNum != 0 )
fprintf( writePtr, "%-6d%-16s%-11s%10.2f\n",
client.acctNum, client.lastName,
```

```
client.firstName, client.balance );
}
fclose( writePtr );
}
}
void updateRecord( FILE *fPtr )
{
int account;
double transaction;
struct clientData client = { 0, "", "", 0.0 };
printf( "Enter account to update ( 1 - 100 ): " );
scanf( "%d", &account );
fseek( fPtr, ( account - 1 ) * sizeof( struct clientData ),
SEEK_SET );
fread( &client, sizeof( struct clientData ), 1, fPtr );
 if ( client.acctNum == 0 )
printf( "Acount #%d has no information.\n", account );
else {
printf( "%-6d%-16s%-11s%10.2f\n\n",
client.acctNum, client.lastName,
client.firstName, client.balance );
printf( "Enter charge ( + ) or payment ( - ): " );
scanf( "%lf", &transaction );
client.balance += transaction;
printf( "%-6d%-16s%-11s%10.2f\n",
client.acctNum, client.lastName,
client.firstName, client.balance );
fseek( fPtr,
( account - 1 ) * sizeof( struct clientData ),
SEEK_SET );
fwrite( &client, sizeof( struct clientData ), 1,
fPtr );
}
}
void deleteRecord( FILE *fPtr )
{
```

```
struct clientData client,
blankClient = { 0, "", "", 0 };
int accountNum;
printf( "Enter account number to "
 "delete ( 1 - 100 ): " );
scanf( "%d", &accountNum );
fseek( fPtr,
( accountNum - 1 ) * sizeof( struct clientData ),
SEEK_SET );
fread( &client, sizeof( struct clientData ), 1, fPtr );
 if ( client.acctNum == 0 )
printf( "Account %d does not exist.\n", accountNum );
else {
fseek( fPtr,
( accountNum - 1 ) * sizeof( struct clientData ),
 SEEK_SET );
fwrite( &blankClient,
sizeof( struct clientData ), 1, fPtr );
}
}
void newRecord( FILE *fPtr )
{
struct clientData client = { 0, "", "", 0.0 };
int accountNum;
printf( "Enter new account number ( 1 - 100 ): " );
scanf( "%d", &accountNum );
fseek( fPtr,
( accountNum - 1 ) * sizeof( struct clientData ),
SEEK_SET );
fread( &client, sizeof( struct clientData ), 1, fPtr );

if ( client.acctNum != 0 )
printf( "Account #%d already contains information.\n",
client.acctNum );
else {
printf( "Enter lastname, firstname, balance\n? " );
scanf( "%s%s%lf", &client.lastName, &client.firstName,
```

```
&client.balance ); client.acctNum
= accountNum; fseek( fPtr, (
client.acctNum - 1 ) *
sizeof( struct clientData ), SEEK_SET );
fwrite( &client,
sizeof( struct clientData ), 1, fPtr );
}
}
int enterChoice( void )
{
int menuChoice;
printf( "\nEnter your choice\n"
"1 - store a formatted text file of acounts called\n"
"\"accounts.txt\" for printing\n"
"2 - update an account\n"
"3 - add a new account\n"
"4 - delete an account\n"
"5 - end program\n? " );
scanf( "%d", &menuChoice );
return menuChoice;
}
```

Output

```
After choosing option 1 accounts.txt contains:
AcctLast Name First NameBalance
29Brown Nancy-24.54
33DunnStacey 314.33
37BarkerDoug 0.00
88Smith Dave 258.34
96Stone Sam 34.98

Enter account to update (1 - 100): 37
37BarkerDoug 0.00
 Enter charge (+) or payment (-): +87.99
37BarkerDoug87.99
Enter new account number (1 - 100): 22
Enter lastname, firstname, balance
? Johnston Sarah 247.45
```

5.7. Command Line Arguments

It is possible to pass some values from the command line to your C programs when they are executed. These values are called command line arguments and many times they are important for your program specially when you want to control your program from outside instead of hard coding those values inside the code.

The command line arguments are handled using main() function arguments where argc refers to the number of arguments passed, and argv[] is a pointer array which points to each argument passed to the program.

The two dummy arguments to the main function are called argc and argv.

- argc contains the number of command-line arguments passed to the main program and
- argv[] is an array of pointers-to-char, each element of which points to a passed command-line argument.

Following is a simple example which checks if there is any argument supplied from the command line and take action accordingly:

```
#include <stdio.h>
int main( int argc, char *argv[] )
{
 if( argc == 2 )
 {
printf("The argument supplied is %s\n", argv[1]);
 }
 else if( argc > 2 )
 {
printf("Too many arguments supplied.\n");
 }
 else
 {
printf("One argument expected.\n");
 }
}
```

When the above code is compiled and executed with a single argument, it produces the following result.

```
$./a.out testing
The argument supplied is testing
```

When the above code is compiled and executed with a two arguments, it produces the following result.

```
$./a.out testing1 testing2
Too many arguments supplied.
```

When the above code is compiled and executed without passing any argument, it produces the following result.

```
$./a.out
One argument expected
```

It should be noted that argv[0] holds the name of the program itself and argv[1] is a pointer to the first command line argument supplied, and *argv[n] is the last argument. If no arguments are supplied, argc will be one, otherwise and if you pass one argument then argc is set at 2.

You pass all the command line arguments separated by a space, but if argument itself has a space then you can pass such arguments by putting them inside double quotes "" or single quotes ''. Let us re-write above example once again where we will print program name and we also pass a command line argument by putting inside double quotes:

```
#include <stdio.h>
int main( int argc, char *argv[] )
{
 printf("Program name %s\n", argv[0]);

 if( argc == 2 )
 {
printf("The argument supplied is %s\n", argv[1]);
 }
 else if( argc > 2 )
 {
printf("Too many arguments supplied.\n");
 }
```

```
else
{
printf("One argument expected.\n");
}
}
```

When the above code is compiled and executed with a single argument separated by space but inside double quotes, it produces the following result.

```
$./a.out "testing1 testing2"
Progranm name ./a.out
The argument supplied is testing1 testing2
```

A simple example follows, which checks to see if only a single argument is supplied on the command line when the program is invoked

```
#include <stdio.h>
main(int argc, char *argv[]) {
if (argc == 2)
printf("The argument supplied is %s\n", argv[1]);
else if (argc > 2)
printf("Too many arguments supplied.\n");
else
printf("One argument expected.\n");
}
```

Note that *argv[0] is the program name itself, which means that

- *argv[1] is a pointer to the first –actual‖ argument supplied, and
- *argv[n] is the last argument. If no arguments are supplied, argc will be one. Thus for n arguments, argc will be equal to n+1.

Example

```
#include <stdio.h>
int main ( int argc, char *argv[] )
{
if ( argc != 2 ) /* argc should be 2 for correct execution */
{
```

```
/* We print argv[0] assuming it is the program name */
printf( "usage: %s filename", argv[0] );
}
else
{
// We assume argv[1] is a filename to open
FILE *file = fopen( argv[1], "r" );
/* fopen returns 0, the NULL pointer, on failure */
if ( file == 0 )
{
printf( "Could not open file\n" );
}
else
{
int x;
/* read one character at a time from file, stopping at EOF, which
 indicates the end of the file.Note that the idiom of "assign
 to a variable, check the value" used below works because
 the assignment statement evaluates to the value assigned. */
while( ( x = fgetc( file ) ) != EOF )
{
printf( "%c", x );
}
fclose( file );
}
}
}
```

This program is fairly short, but it incorporates the full version of main and even performs a useful function. It first checks to ensure the user added the second argument, theoretically a file name. The program then checks to see if the file is valid by trying to open it. This is a standard operation, and if it results in the file being opened, then the return value of fopen will be a valid FILE*; otherwise, it will be 0, the NULL pointer. After that, we just execute a loop to print out one character at a time from the file. The code is self-explanatory, but is littered with comments

5.7.1. Files Examples

1. **Write a C program to read name and marks of n number of students from user and store them in a file.**

```
#include <stdio.h>
#include<conio.h>
int main(){
 char name[50];
 int marks,i,n;
 clrscr();
 printf("Enter number of students: ");
 scanf("%d",&n); FILE *fptr;
 fptr=(fopen("C:\\student.txt","w"));
 if(fptr==NULL){
 printf("Error!");
 exit(1);
 }
 for(i=0;i<n;++i)
 {
printf("For student%d\nEnter name: ",i+1);
scanf("%s",name);
printf("Enter marks: ");
scanf("%d",&marks);
fprintf(fptr,"\nName: %s \nMarks=%d \n",name,marks);
 }
 fclose(fptr);
 getch();
 return 0;
}
```

2. **Write a C program to read name and marks of n number of students from user and store them in a file. If the file previously exits, add the information of n students.**

```
#include <stdio.h>
#include<conio.h>
int main(){
 char name[50];
```

```
int marks,i,n;
clrscr();
printf("Enter number of students: ");
scanf("%d",&n); FILE *fptr;
fptr=(fopen("C:\\student.txt","a"));
if(fptr==NULL){
printf("Error!");
exit(1);
}
for(i=0;i<n;++i)
{
printf("For student%d\nEnter name: ",i+1);
scanf("%s",name);
printf("Enter marks: ");
scanf("%d",&marks);
fprintf(fptr,"\nName: %s \nMarks=%d \n",name,marks);
}
fclose(fptr);
getch();
return 0;
}
```

3. **Write a C program to write all the members of an array of structures to a file using fwrite(). Read the array from the file and display on the screen.**

```
#include<stdio.h>
#include<conio.h>
struct s
{
char name[50];
int height;
};
int main(){
struct s a[5],b[5];
FILE *fptr;
int i; clrscr();
fptr=fopen("file.txt","wb");
```

```
for(i=0;i<5;++i)
{ fflush(stdin);
printf("Enter name: ");
gets(a[i].name);
printf("Enter height: ");
scanf("%d",&a[i].height);
}
fwrite(a,sizeof(a),1,fptr);
fclose(fptr);
fptr=fopen("file.txt","rb");
fread(b,sizeof(b),1,fptr);
for(i=0;i<5;++i)
{printf("Name: %s\nHeight: %d",b[i].name,b[i].height);
}
fclose(fptr);
getch();
}
```